D1429385

The Emerging Global Food System

The Emerging Global Food System

Public and Private Sector Issues

Edited By

GERALD E. GAULL
Director, Center for Food and Nutrition Policy and The Ceres Forum
Georgetown University

RAY A. GOLDBERG
Moffett Professor of Agriculture and Business
Graduate School of Business Administration
Harvard University

JOHN WILEY & SONS, INC.
NEW YORK / CHICHESTER / BRISBANE / TORONTO / SINGAPORE

This text is printed on acid-free paper.

Library of Congress Cataloging in Publication Data:

The emerging global food system : public and private sector issues / edited by Gerald E. Gaull, Ray A. Goldberg.
p. cm.
Includes index.
ISBN 0-471-59072-X
1. Nutrition policy—Congresses. 2. Food industry and trade—Congresses. 3. Food supply—Congresses. I. Gaull, Gerald E. II. Goldberg, Ray Allan, 1926-
TX359.E46 1993
363.8'56—dc20 93-16589
CIP

Printed in the United States of America

10 9 8 7 6 5 4 3 2

Contents

Acknowledgments

We are grateful to Mr. Robert Flynn, Chariman of the Board and Chief Executive Officer, the NutraSweet Company, for making the second Ceres Conference possible and to Mr. Max Downham, Vice-President, Strategy and Planning, who was responsible for so much of its execution. The participants, both from the public and the private sectors, provided the papers and the provocative discussion that made the event so innervating.

Mr. Eric Marx provided excellent editorial assistance and Misses Cristi Arbuckle and Barbara Ashley typed and formatted the manuscript disk.

Foreword

John T. Dunlop

This volume, and the second Ceres Conference of October 30–November 1, 1991 that it reports, portrays the global food system in rapid transition in all of its features—the biological revolution and food technology, the potential of nutrition, the new forms and scope of business organizations, the diverse regulatory processes and their harmonization, the standards for safety and risk, the conflicts over international trade policies, including the relations between developing and advanced countries, and the public willingness to validate these often discordant changes.

The global food system, in the past decade particularly, has been undergoing a restructuring through the consolidation and internationalization of enterprises at every stage. Horizontal and vertical coordination has been taking place through joint ventures, partnering, and long-term procurement and marketing arrangements. These changes have been abetted by the technological and information revolutions. Supplies to farms are now provided by fully technological companies; commodity processors of traditional products have been transformed into global wholesalers of custom-designed food, feed, fuel, and pharmaceutical ingredients; and consumer market enterprises are providing full consumer service and logistic systems.

The rapid technological changes and global business restructuring have understandably created tensions and conflicts within the established national research and regulatory processes and agencies designed to deal with agricultural marketing, standards for exports and imports, food safety, environmental protection, and the transition from research to commercial use. As the editors state in their conclusion, "What is good for the world food system in the long run is often disruptive to national systems in the short term...." The diverse agencies and policies have not always been congruous, coordinated or timely in their decisions. The litigious quality of regulations has enhanced these difficulties. To deal with these issues, new international and regional agencies are arising, for example, in the Uruguay Round of the GATT as related to agriculture and in the European Community that seeks to develop by 1993 a common food policy that will lay down common safety requirements. (Caroline Jackson's discussion of harmonization and hurdles in Part II is particularly informative.)

In its international dimensions the food system is confronted by the insistent enigma of both surpluses and hunger. The reorganization of agriculture and food production and distribution in Eastern Europe and the republics of the former Soviet Union, the North American trade treaty, changes in the European Community agricultural policies, and famine in parts of the Third World all provide an indication of the forces generating further transformation in the global food system.

A particularly instructive session in Ceres II was provided by the Public Policy Panel on Food Safety with Charles R. Nesson as moderator (Chapter 18). The panel included participants from technologically innovative companies, the scientific community, regulatory agencies, elected national legislators, the press, public relations officers, and consumer, environmental, and public interest groups. These panel exchanges, a discourse with the full Conference membership, well illustrated the range and intensity of conflicts and tensions inherent in biotechnological changes in the food system. The standards of risk assessment and safety, the regulatory process, the complexities of reactions among diverse groups in the general public, the role of media, and the activities of public interest groups were all vividly reflected in the session.

The discussion also illustrated the potential of sustained and frank discourse for reducing conflict and building consensus. The creditability of the new food system requires such consensus building to legitimate its processes with the public and to speed the introduction of acceptable change.

Ceres II underscored the necessity for a continuing forum of leaders drawn from major sectors of the global food system: representatives of the scientific community, business organizations, the regulatory and political processes, and consumer and environmental groups, such as assembled at this Conference. Exchanges of sharp views at long range through the media or at legislative hearings are not a substitute. Such confrontational processes alone do not well serve, moreover, the global nature of the production and distribution side of the food system and its regulation.

A greater degree of consensus in such a forum is possible, in my experience, on questions of standards or processes for an acceptable degree of safety and risk with the introduction of new products. It should also be possible to develop a greater degree of consensus on ways to harmonize the growing international regulation of the food chain.

Further, new processes of regulation, such as negotiated rulemaking in the United States (Negotiated Rulemaking Act of 1990), seek involvement of all interested parties to develop consensual rules. These procedures reduce adversarial relations, minimize litigation, and will help to speed the potentials of biotechnological change.

For such a continuing forum to achieve its objectives, experience teaches that preliminary work by staff, drawn from the diverse groups, is requisite to fruitful periodic discourse among principals. The joint gathering and review of data, preliminary appraisals, and identification of key issues for discussion are essential facilitating steps.

Since persistent change is certain for the global food system, this Ceres Conference reached the conclusion that a continuing global forum of representative leaders should be established to develop greater creditability and consensus on the issues arising over the introduction of biotechnological change and the associated regulatory processes.

Introduction: National Food Policies in a Global Economy

The Honorable Roger B. Porter

THE QUEST FOR EFFICIENT FOOD PRODUCTION

Although the invention of the airplane and the advent of television often attract more attention, one of the most dramatic changes of the 20th century has been the revolution in agriculture. Early in this century, some were concerned about whether or not we could grow enough food to supply the needs of a growing population around the world. Thanks to technological advances, we can now produce all the food we need and more.

The number of acres under cultivation in the United States—roughly 330 million—has remained almost unchanged since the beginning of the century. During that same period, our farm population declined from 30 million to 5 million, from 40% to less than 2% of our population. This decline in our farm population has occurred gradually and with relatively little social friction.

Advances in science and technology have made each farmer dramatically more productive. In 1930 it took a farmer six to eight hours to produce an acre of corn, using horses and early tractors. Only two to three hours is required to accomplish the same task today, using modern tractors and harvesting equipment. In 1940 each farmer fed 18 1/2 people. Today he feeds 128 people: 94 in the United States and 34 abroad. Indeed, since the end of the Second World War, the average yield per acre in the United States for all crops has more than doubled.

We have become so successful in increasing our supply of food that the principal challenge confronting us is one of distribution. How do we get food to those who need it? We face this problem on a large scale today in the Soviet Union. United States Department of Agriculture officials, trying to help find a solution to this problem, returned from the Soviet Union recently with stories of extraordinary waste. For example, it is estimated that nearly half of the 1991 potato crop will have rotted on its way to market.

Other countries have also taken advantage of this progress in agricultural science and technology. Today worldwide capacity to grow food has greatly outstripped demand. India is now a net exporter of grains, a situation unthinkable a quarter of a century ago. Even China is an occasional net exporter of grain.

Technology will almost certainly continue to produce increased yields and enhanced supply. Moreover, there is no reason to believe that the increase in demand, due to population growth, will exceed the growth in supply due to technological advances and their diffusion around the world. In short, global agriculture is likely to become even more competitive rather than less competitive.

As supply has outstripped demand, countries have acted predictably. Where there is an imbalance in supply and demand, it is natural for individuals, firms, and nations to attempt to protect themselves and their market share. Nearly every nation, including the United States, has put in place measures designed to protect its domestic producers from international competition. These measures are costly. According to the OECD's most recent estimate, the total cost of agricultural support to taxpayers and consumers, just in OECD countries, exceeds $245 billion annually. This is a heavy price to pay in terms of economic inefficiency.

Over the long term, a persistent and undiminished imbalance between supply and demand is unacceptable. Government officials in other countries acknowledge that such inefficiency is unacceptable. No one defends the current arrangements as optimal. Virtually everyone agrees that supply and demand must be brought into closer balance and that the way to achieve such a balance is through greater reliance on market forces. But fundamental change is never easy in any country or internationally, and precipitous change inevitably brings with it a justifiable charge that one group or interest is being treated unfairly.

In shifting from our current arrangements to arrangements that will bring supply and demand into closer balance, we face two central tasks. The first concerns the pace of change. No one has suggested an immediate total reliance on the marketplace for the production of agricultural goods. There must be a transition period for needed human and economic adjustments to take place. It is crucial that the adjustments actually occur and that countries not use the need for a transition period as an excuse to forego making the necessary changes. We must firmly place policy on a trajectory that will take us where we want to go.

The second task concerns who is going to bear the burden of the needed adjustments, which will vary from country to country. These adjustments must reflect the social and political realities that exist in individual nations. This is nothing new. The adjustments that have occurred in agricultural production and employment in the United States and other industrialized economies during this century are ample testimony to the capacity of nations to accommodate successfully the changes wrought by technology and increased global competition.

Both of these dimensions have captured the agenda for the Uruguay Round of trade negotiations. For the past five years, negotiators from the United States and approximately 100 other nations have been involved in the most comprehensive trade negotiations ever undertaken. This sweeping initiative, which is being conducted under the General Agreement on Tariffs and Trade (GATT), has as its basic objective to modernize and to enhance its utility for the remainder of this decade and into the next century. While the negotiations include the traditional subjects of tariffs, subsidies, and dumping, rules are also being developed in three areas of special interest to us.

First, the negotiators are attempting to strengthen today's inadequate rules with respect to trade in agricultural products. The current GATT rules, for example, do not set any meaningful limits on agricultural export subsidies, on trade-distorting domestic support policies, or on import access restrictions. As a result, trade in agricultural commodities today more often reflects government's ability to buy market share than producer's ability to compete.

Second, the Uruguay Round negotiators are developing rules to protect intellectual property rights. During the course of the negotiations, the views of many Third World countries have changed quite dramatically. In 1986, when the round began, a large number of developing countries saw intellectual property rights as a conspiracy by wealthy countries to keep the rest of the world in a permanent state of dependency. A remarkable change has occurred in the last five years. Most developing countries now recognize that without adequate patent and trademark protection they will be unable to attract the foreign investment that they desire in high technology industries. The biotechnology industry clearly requires firm rules on intellectual property if it is to flourish.

A third area of special interest is the effort to require basing health-related standards for food and agriculture products on scientific evidence, so that these standards can not be used as disguised barriers to trade. The current GATT rules recognize that member countries have a right to adopt health measures deemed necessary to protect human, animal, or plant life. The rules are so vague, however, that countries often use health requirements to disguise barriers to trade that have little, if anything, to do with health protection. Although the specific details remain to be worked out, it appears that we are moving toward agreement in the Uruguay Round that all countries would have the right to set standards stricter than international standards. The scientific foundation of national regulations that are more stringent than the recognized international standard, however, could be challenged by an exporting country under new dispute settlement procedures.

The Uruguay Round has made limited progress since December 1990, when a ministerial level negotiating session in Brussels broke down over the issue of agriculture. The central problem was the European Community's inability or unwillingness to provide its negotiators with a mandate to negotiate reforms relating to export subsidies, domestic price supports, and market access.

There are now some encouraging signs that these roadblocks are being cleared away. The German cabinet and chancellor have announced a commitment to seek a compromise on agriculture in order to make the Uruguay Round succeed. Other EC members, including France, appear willing to consider some changes in their agricultural policies. There are many obstacles to be overcome in the Uruguay round, but the outlook is more optimistic now than it has been for a long time. We should know by the end of 1991 whether agreement will be possible.

NATIONAL FOOD SAFETY POLICIES

The Uruguay Round debate on international standards has some immediate implications for citizens in all countries because governments around the world recognize that consumers want a plentiful supply of safe, affordable, nutritious food. Public policies designed to foster such a food supply are, of necessity, a mixture of science and economics, culture, and geography. Policymakers can objectively consider these factors. Yet, to produce a politically acceptable and workable system, the public needs to understand each of these elements and the results that will flow from according different weights to competing considerations.

For example, the policy debate in many developed countries surrounding the desire for a risk-free food supply is a luxury that many developing countries, more concerned with simply providing an adequate supply of nutritious food, cannot yet afford. Each society has to judge for itself where the proper balance lies between these two objectives. Those decisions will reflect a nation's internal policies and its external trading position.

In the United States, the threshold question often asked by opponents of food safety legislation is: Why do we need to legislate at all? Certainly there are theoretical risks associated with the use of pesticides on food, but there are risks everywhere in our society. One of the great challenges and responsibilities of policymakers in an era of advancing technology and communication is determining what risks government ought to seek to eliminate or reduce, what theoretical risks are acceptable, and whether or not, in eliminating some risks, we are creating others.

"Is eliminating the theoretical risk worth the price?" is a question implicit in every discussion of risk. For example, it is theoretically possible to design a car that would protect the driver and passengers from virtually all injury. But the cost would be exorbitant and very few people could afford such a vehicle. We do not have any national policy requiring that cars be designed to provide that level of injury protection. Accordingly, an American has about 18 chances in 100,000 every year of dying in an automobile accident. Likewise, those who have smoked for 20 years or more have a 700 in 100,000 chance annually of dying because of their smoking habit. What about people who continue to eat large quantities of eggs? What is the increased chance of heart disease associated with a country breakfast every morning?

We have struggled to make cars safer. We have limited the exposure of non-smokers to smoke. We have tried to educate people about the risks of cigarettes and of high cholesterol. But we have not prohibited the sale of cars, cigarettes, or eggs.

Why then, in the case of carcinogenic pesticides with risk magnitudes far less than those associated with cars, cigarettes, or eggs, do consumers and environmentalists ask us to do more? Why are we asked to draw a bright line at an absolute maximum risk of one in a million over a 70-year life expectancy, regardless of the benefits obtained at a slightly higher risk level?

In my view there are two reasons why we have established a different and higher safety standard for pesticides than for cars, cigarettes, and eggs. First, the risks associated with cars, cigarettes, or eggs are readily identifiable, permitting individuals to make informed choices. Putting aside the issue of passive smoking, I can choose to smoke or not to smoke. I can choose to drive or to take some other form of transportation. In other words, I have some appreciation of the risk. I know it when I see it, and I can choose, if I wish, to avoid it.

But what about the food that I eat? Currently, pesticides are key to increased yields and, as importantly, in preserving food until it gets to the marketplace. But none of us has the power of eyesight or of intuition necessary to determine what amount of residue, if any, of a particular pesticide remains on the piece of fruit we purchase. To a large extent, since we all have to eat, we are left without meaningful choice. Some people avoid the risk of pesticides by buying only organically grown foods. But could we all? And if we did, would we have enough food of suitable nutritional value?

Second, because consumers are unable to identify this risk and to avoid it routinely, government has a particularly important role in assessing and educating the public about the risk. We haven't always met this challenge well. Perhaps the stiffest challenge to policymakers in all areas of regulation is combating one of the unwelcome results of technological advancement: misinformation can be dispersed as quickly as accurate information.

For example, not too long ago, a television program presented unsubstantiated and hyperbolic statements about the safety of a commonly used pesticide, Alar, on apples. Consumer reaction to the story severely crippled the apple industry in large parts of the United States. It was not until the well-publicized voluntary removal of the pesticide for use on apples occurred that the anxiety of the American public began to subside.

In that case, the Federal Government did a poor job of assuaging the fears of the public. We neither adequately educated consumers about the facts nor conveyed the credibility needed to overcome the sense of anxiety that raced through the public.

In this area, credible assessments of risk and enhancing consumer confidence in the pronouncements of regulators needed to be buttressed by those in the scientific community who could provide that credibility. A well-intentioned and well-informed assistant administrator is no match for a movie star or an activist when it comes to getting on the evening news and discussing the risks and benefits of chemicals or processes that affect our food supply.

There is some reason to be optimistic on this front. The defeat of the Big Green Initiative in California in 1991 provides at least circumstantial evidence that our goal may not be totally illusive. But we have a great deal more to do.

FOOD SAFETY AND THE DELANEY PARADOX

Notwithstanding these reasons for tackling the food safety issue, however, some criticize current reform efforts as unnecessary because a comprehensive regulatory scheme already exists. A brief history of the so-called Delaney Clause is instructive.

The Federal Food, Drug, and Cosmetic Act regulates the presence of pesticides on food in two ways. First, the Environmental Protection Agency sets tolerances for the level of pesticides that are allowed on raw agricultural products. These tolerances are to be set at levels that are necessary to protect public health, while taking into account the need for an adequate, wholesome, and economical food supply. In short, they are to try to balance competing interests.

By contrast, the Environmental Protection Agency (EPA) may not grant a tolerance for the presence of a pesticide residue in processed foods if the pesticide has been found to induce cancer. The Delaney Clause allows no carcinogenic pesticide, no matter how weak or how great the benefits, to be used if any residue of the pesticide would remain in processed foods. As a result, a carcinogenic pesticide may be used on a raw tomato, creating at most a negligible risk, but that same tomato cannot be used for processed spaghetti sauce.

As troubling as this disparate treatment was when the Delaney Clause was first enacted in 1958, its application was minimized, at least in part, by our relative inability to detect residues at that time. But as technology has advanced, our ability to measure residues has increased by orders of magnitude, exacerbating this disparity.

The EPA is now applying the Delaney Clause standard to new pesticides for which registration is sought, but not to the older and generally less safe pesticides. This phenomenon might appropriately be called the Delaney Paradox. Through improved toxicological measurement methods, the Delaney Clause, which was originally intended to promote safety, has, in fact, become a barrier to bringing new and safer pesticides on the market because they are required to meet a wholly different standard than the existing pesticides they are seeking to replace.

The 1988 amendments to the Federal Insecticide, Fungicide, and Rodenticide Act (FIFRA) require reconsidering all older pesticides in light of contemporary science. As older pesticides up for reregistration are canceled as unsafe, the Delaney Paradox threatens to eliminate some of the pesticides that make possible broad segments of the U.S. food industry. Without an adequate standard, there well may be no adequate substitutes.

The pending food safety legislation debate presents a welcome opportunity for the Federal Government to design a rational framework for making assessments of acceptable risk and to instill new confidence in the public that consumers will not be exposed to unacceptable risk levels. It will permit us to anticipate, rather than merely to be reactive to, the public's concerns.

The Bush administration advanced four principal considerations in its food safety legislation: first, the desire to eliminate what I have described as the Delaney Paradox; second, the need to streamline the suspension and cancellation procedures under FIFRA; third, a desire to produce nationally uniform standards for goods that are in interstate commerce, so that business doesn't have to respond to 50 different masters as opposed to one; and fourth, an effort to enhance consumer confidence in the safety of our food supply and to avoid the unnecessary disruptions that have occurred in the past because of scares, panics, and fears that have arisen, largely through the media.

REGULATORY REASONABLENESS IN FOOD SAFETY

There was general agreement between Congress and the Bush administration on the need to move to a standard of negligible risk. We view negligible risk as approximately one in a million. Others prefer a "bright line" of one in a million. More importantly, some are proposing never crossing that line: not to prevent a greater health hazard, as in the case of aflatoxin; not to prevent food shortages or great price increases; not even to suspend it temporarily to prevent economic ruin to growers. The Bush administration felt strongly that regulators confronting natural risks, man-made risks, and unforeseen circumstances must have the flexibility to consider such benefits in determining whether or not and when to exceed the agreed risk level.

In short, we need to adopt a regime of regulatory reasonableness. We believe that if we do our job properly with the help of those in the scientific community, then the public will permit adequate and appropriate flexibility for regulators. We have a responsibility to ensure that if an absolute limit is what the public demands, then the cost of that "bright line" and the relationship between the risks and the benefits of breaching it are understood.

Another crucial issue is whether or not we should adopt a uniform national standard with the opportunity for only limited exceptions. The issue of uniformity has important implications for both our national commerce and our international trade agreements. A lack of uniformity in which each state has the ability to set its own risk levels not only creates potentially significant problems for those who ship commodities in interstate commerce, but it is also inconsistent with our ongoing efforts to harmonize these standards through the GATT negotiations. In this important respect national uniformity makes sense.

EMERGING TECHNOLOGIES

Finally, the evidence suggests that enormous potential exists for substantial developments in biotechnology. Biotechnology is still at a very early stage in its development. Awe and wonder shape our vision of the future. Vannevar Bush, President Truman's science adviser, spoke of observing the endless frontier, a phrase he chose as the subtitle of the book based on his report to President Truman on science in the post-World War II period. At that time, the term "biotechnology" was virtually unknown, and federal investment in the fields that would become biotechnology focused on the biological sciences.

The connection to the biological sciences, particularly as they relate to human life and disease, has been at the core of the federal government's involvement in biotechnology. In 1991 the President proposed total federal investment in biotechnology of more than $4 billion, of which about $3 1/2 billion is directed to the Department of Health and Human Services. The jewel in the crown of U.S. biomedical research is the human genome project, an effort to produce a chromosome map within five years and a complete map of all human DNA in 15 years.

Interestingly the Japanese science community has been piqued by interest in another genome, rice. The first priority for Japanese researchers is to increase the nutritional value and the productivity of that important food.

A look at the endless frontier of biotechnology reveals several recognizable political interests and forces at work. First, some fear, or will play on, fears of the unknown. This problem is inherent in pushing forward any new frontier. Second, bureaucracies at both ends of Pennsylvania Avenue, in the Executive Branch and in the legislature, will engage in fierce competition for regulatory control of biotechnology. If this bureaucratic competition is not appropriately channeled, it could lead to excessive regulation. Third, some seek to limit biotechnological advances out of a fear of increased competition—for example, those who would seek to bar the use of bioengineered substances. This may well prove the most important international dimension and possible barrier to technological progress. It is one of the fears that many U.S. biotechnology companies face.

In August 1990, President Bush approved four principles to guide federal regulatory behavior with respect to our development of biotechnology. First, oversight should focus on the characteristics and risks of the biotechnology product, not on the process by which it was created. In other words, a concentration on ends and not on means. Second, federal review should seek to minimize regulatory burdens while assuring the protection of public health and welfare. Third, regulatory activity should be designed to accommodate rapid advances in biotechnology. Therefore, performance-based standards are generally preferable to design standards. Finally, to create opportunities for applying innovative biotechnological products, all environmental and health regulation should, wherever possible, use performance standards.

Stimulating development in biotechnology requires intellectual property protection. The effort to assure this protection is centered in the Uruguay Round of GATT negotiations. If these negotiations succeed, they will help move us toward a regime of regulatory reasonableness in which biotechnology is likely to flourish. We must strive for regulatory certainty. One of the most difficult challenges to those on the frontier of scientific innovation is anxiety about what they must go through, once they produce a breakthrough, in order to get it adopted and commercialized.

We must also continue our commitment to basic research. Such research holds the promise of long-term gains to society. This is an investment in the future that is worth making.

CONCLUSION

These are exciting times, with both great opportunity and great challenges. We can realize enormous continued gains in the efficiency of agricultural production, but this will require us to adopt trading arrangements that will, of necessity, involve adjustments. We must have both the political will to make those adjustments and the wisdom to ensure that the burdens associated with those adjustments are distributed in a fair and equitable way.

We can also move into an era of greater safety with respect to our food supply. We have been on this path for some time. The safety of our food supply has improved, and we can continue to make additional improvements. Our challenge as a society is to define what are acceptable risks, to monitor compliance successfully, and to communicate this in a way that the public is reassured that the food supply is safe. Finally, in advancing the frontier of biotechnology, we need to encourage and guide innovation without stifling it by a regulatory structure that overwhelms and smothers it.

This is an area of public policy where optimism is warranted. If we act wisely, we can achieve not only greater efficiencies in the way we produce and distribute food, but also a safer food supply and remarkable advances in biotechnology. It will take much wisdom and will. But it is an achievement within our grasp.

DISCUSSION PORTER:

DR. KORNBERG: In understanding carcinogenesis, we should recognize that the assays for carcinogenicity are still very primitive. According to my colleague, Bruce Ames, much of what we eat contains naturally occurring carcinogens in amounts that far exceed those pesticides that have been identified as putative carcinogens, some by orders of magnitude. The admonition to avoid eating a food exposed to a pesticide, thus, would need to apply to much of the food supply.

The second concern I have is that the government agencies under the present administration are not behaving in the way that you describe. The FDA, for example, has a formidable bottleneck that actually sabotages the rapid development and assessment of new technologies. The EPA, by setting standards that are so difficult to meet, such as 99% effectiveness, does so at a cost that makes the whole process unreasonable. As a result, important problems are completely ignored.

I wonder whether you can bring us some ray of hope that the government itself will put its house in order.

DR. PORTER: I don't know whether this should be construed as optimism, but let me share with you an experience that confirms at least in part the point that you are making.

I gathered in my office for several sessions the head of the Food and Drug Administration, the Assistant Administrator responsible for food regulation at the Environmental Protection Agency, and the Deputy Secretary of Agriculture to discuss trying to develop a coordinated administration position on food safety.

We got into endless discussions of what I have described as a bright line, the one in a million chance that you could die of cancer because you were eating a carcinogen. At one point I turned to the FDA Commissioner and said, "What are my chances of dying of cancer if I eat only organically grown foods? If I never consumed anything that had a pesticide on it, what would be my chances of dying of cancer?" He said, "If you smoke, your chances of dying of cancer would be about 25 in a hundred. In the United States, over a fifth of the people who die of cancer die of smoking-related instances. It is the number one killer."

He said, "If you don't smoke and you didn't eat any foods with pesticides on them, the fact that you have to eat in order to stay alive, means that your chances of dying of cancer are 7 in a hundred." I said, "Let me make sure I have this straight. My chances of dying of cancer if I don't eat food with a given pesticide on it are 70,000 in a million, and we are arguing about the difference between 70,001 in a million and 70,000 in a million?" And he said "You got it."

I then turned to the assembled audience and I said, "I think this is insane. I know from personal experience how frustrated you are with the behavior of these bureaucracies." It is a painful process to turn around a bureaucracy. Turning around FDA and

EPA and the regulators at USDA is going to take more than the efforts of just this administration. It is going to take a persistent, sustained effort over time.

This is the first time, I am told, that we have gotten all the regulators into one room and started the process of knocking heads together. It has been a good experience for those of us who were firing the shots at them. At least it was cathartic.

The more we can get the message out about what kind of risks we are really talking about, the more we can bring public pressure to bear on the kind of behavior that you are talking about. We have got to adopt a regulatory regime that does not stifle innovation and creativity. The talent in our society is going to flow to those areas where it thinks it can make the greatest contribution. If it gets into an area where all it finds are roadblocks, it will pick up and go somewhere else. The most highly regulated industries in the United States are not the industries that show the greatest progress. Why? Because the talent tends to flow to where it thinks it can have the greatest impact. If we want to see a lot of progress and breakthroughs in biotechnology, then we have got to create an environment and a regime in which that is going to happen.

DR. MILLER: As a former member of this lethargic and incompetent bureaucracy, I feel compelled to point out a couple of things. First of all, this is not the first time that these agencies have gotten together in a room to discuss these issues.

The first time was probably long before I arrived in government in 1978. The problem that bureaucracies face is that their lords and masters set down the standards under which they must operate. However those standards are interpreted by the Congress and the political administration; that is how the bureaucrats enforce it.

A much more important issue concerns the question of rational regulation. If a regulator is uncertain of the facts of a situation or does not feel competent to defend a judgment in that situation, the regulator will say "no." The quickest way for a gatekeeper to close the gate is to become frightened, and the fact of the matter is that our regulatory agencies have become more and more unsure of themselves in this era.

One of the reasons this has happened is that many of the people who were extremely competent in the regulatory business have left government. Neither this administration nor the two previous ones have provided the regulatory agencies with the resources they need to regulate this gigantic industry, the food industry. As a result, many very competent people left. There were many competent people in government who were there largely because they felt they were doing some good. Many of us left at the same time, when it became clear that what was left of the agency was becoming a hollow shell. Year after year, FDA officials were instructed not to ask for new funds, to maintain the fiction that they could do with what they had. And all the while they were burning the supports that maintained them.

Dr. Kessler is trying to rebuild, but he has still not managed to provide resources to develop the best possible people in the regulatory agency. If you are going to regulate biotechnology, you need to have people regulating the science who are among the best biotechnologists. You get them by recruiting the best and giving them the best to work with. If you really want good regulators, if you really want a door opened, pay them, support them, and they will be there.

DR. PORTER: Perhaps I misspoke myself by suggesting that that was the only time we had gotten all the regulators in one room. They asserted that it was the only time that we had gotten them in one room and produced a presidential statement with a food safety proposal that all of them had signed on to.

But whatever the facts of previous administrations, you make a very good point. If we are going to have the kind of regulation that we want, we have to attract people with talent into the regulatory agencies and give them the freedom to exercise it. I totally agree with you on that point.

On the matter of resources, we need to recognize the fact that we have a huge imbalance between outlays and revenues right now, as evidenced by continuing budget deficits. If we want to pour more resources into a particular agency we may want to mandate user fees much more extensively. Users may be willing to help in putting up the kind of resources that are needed if they are going to end up getting a regime of regulatory reasonableness.

Part of the problem is at the other end of Pennsylvania Avenue. The fear that many regulators exhibit about exercising more discretion is due to the way they are treated when they get hauled up on the other end of the avenue. Until they start being treated differently in oversight hearings, they are going to behave in very predictable ways. They are going to think "I am going to keep my head down. I am going to hunker down. I am not going to lift my head above the water line because if I do, I may find it getting blown off." There are many people who are willing to leap at the first opportunity to exploit a circumstance in which they think that discretion has been used unwisely. So, we need to do our job in the Executive Branch. We need some cooperation in the Legislative Branch as well.

MR. SALQUIST: You spoke quite clearly and accurately, I believe, in defining the Delaney Paradox, but with reference to your comments to Professor Kornberg, I would like to comment on what I would call the Porter Paradox. That is, you stated in your food agenda that national uniform standards was clearly a priority for the administration and that building consumer confidence in the safety of the food supply was a prime mission of the administration. In fact, the ideologic stance that you described, which is in opposition to any process-based regulation with respect to biotechnology, has exactly the opposite effect from that which you are trying to achieve. In other words, by the federal administration refusing to stand up and adopt a policy that says we are going to pass some judgment on this technology, you are fostering local regulation. You are, in fact, creating a situation in which the consumer does not have confidence in this technology because the administration hasn't spoken up and said "We have looked at it and it is okay."

DR. PORTER: I share your view that if you are going to move to a system of national uniformity and not have a patchwork quilt of standards that are confusing to people, then you are going to have to have a federal government that is willing to step up and say this is acceptable and this is not acceptable. Let me share with you one other experience. I pressed people to tell me the down-side risks of going to a national standard. I asked, "What difference does it make if another state has this or that?" I said, "For example, with the food that I eat, what is the biggest danger?" I asked the head of the FDA, "What would you have me do if you had any incremental monies to spend to protect the food you eat?" He said, "Ninety percent of the problems associated with any food that you eat have nothing whatsoever to do with the standards that are approved with respect to the use of particular pesticides. They have everything to do with the application of those pesticides. It is the farmer who puts too much on his field, doesn't get caught and does it regularly." He said, "If people were really interested in the safety of their food supply, there are a few people misusing the existing pesticides who are a far greater threat to the safety of the food system than the standards that we are spending all this time talking about—one in a million vs. one in ten million, for example. If we had any sense, that is where we would be putting the resources." I have no reason to disagree with his judgment on that matter.

PROFESSOR HIGGINSON: I would like a clarification on one point. You used the word "safety" and it is quite clear that you were referring only to cancer risks. You then used the term "one in a million" as a real risk. Most of us realize that this is a calculation made from certain mathematical models to develop a theoretical risk. It is not a real risk, as widely used in discussion. You then moved to discussion of biotechnology within the same context. This implies that the public will regard biotechnological risks as equivalent to cancer risks. I am sure you are not confusing the two, but I think there is a very great danger that the public will do so.

DR. PORTER: I happened to use that as an illustration because it was the one that I thought would be most familiar and perhaps most interesting. There are a whole host of food safety risks that are not related to carcinogens. You are quite correct that it is a theoretical risk, and there is ample opportunity for the public to be confused about what is going on in their food supply.

Indeed, the scares, panics, and anxieties that we have had in the United States over the last two or three years are ample testimony to this. I might add that it is not simply in the United States. I had the opportunity of having dinner one evening with the minister of agriculture from the United Kingdom, who said that by far the most difficult challenge in his job was dealing with scares and panics about the food supply. He said that if you don't wear a white smock and have a fancy set of initials after your name, your credibility with the public is minimal. He said, "Here we are in the positions of public policymakers and no one there is going to believe us." When we had the scare on Alar, we still had C. Everett Koop as the Surgeon General. He was the one who ended up having to go out and hold a press conference. He was the closest thing we had to somebody in the administration, who we felt could combat the Meryl Streeps and the others who had been out there inducing this sort of panic on the part of the public. The challenge of educating the public about risk is a difficult one.

PART I

THE ROLE OF NEW TECHNOLOGIES IN CHANGING THE GLOBAL FOOD SYSTEM

Ray A. Goldberg

INTRODUCTION

The purpose of this section is to serve as a background for the panelists who, through their private, cooperative research, and governmental institutions, are providing leadership in the global food system. Such leadership enables the use of new technologies to be more responsive to the expanding health and taste needs of society. And it does so in a manner that is economically, environmentally, and structurally sound and with improved safety standards at all stages of the value-added process.

Institutions, functions, coordinating arrangements, markets, technology, and societal value systems are now all changing. These changes are occurring at an ever-increasing pace and magnitude, adding to price and volume volatilities and to a global restructuring of the food system.

This restructuring has occurred through the consolidation of firms at every stage of the value-added food system. In addition horizontal and vertical coordination of the food system is taking place through joint ventures, partnering, and long-term procurement and marketing arrangements. Also, the information revolution, such as the development of scanning devices at the food retailing level, has aided the development of just-in-time inventory arrangements and has provided guidance for the development of both premium supermarket brands and nationally manufactured branded food products. Finally, the biotechnological revolution is taking place at a most advantageous time in both developed and developing food economies. The consolidation, coordination, information and technology breakthroughs that are occurring enable science to respond to specific market and nutritional needs.

DEVELOPED FOOD ECONOMIES

In the case of developed food economies, the ability to tailor-make foods for target markets in terms of taste, nutrition, or other health or style-of-living priorities makes it possible for input suppliers of unique seeds to add their brand to the ultimate product of that particular commodity, for example, Pioneer Hi-Bred International and its "Better Life Foods." Similarly, in the technological breakthroughs in tomatoes, cotton, and canola developed by the Calgene Company, the discovery in each case improved the product and/or the process. Further, fat and sweetener substitutes by Tate & Lyle, NutraSweet, ConAgra, and others enabled the companies to utilize their brands in conjunction with food and beverage manufacturers such as Coca-Cola, Pepsi-Cola, and Kraft General Foods. In other words, these companies became value-added ingredient suppliers that aided manufacturers, food distributors, and drinking and eating establishments to add to the differentiation of their products. In all of these examples, the end-product market was expanded to meet additional nutritional, taste, and style-of-living priorities. In other markets, biotechnological products from agricultural sources have been used to clean up waste disposal sites and to respond to ecological concerns affecting climate and extinctions of species.

DEVELOPING FOOD ECONOMIES

In the developing world the ability to produce additional quantities and improved qualities of food in a leap-frog technological fashion, using familiar crop and animal breeds and also using the existing labor and ingenuity of the small-scale traditional subsistence farmer, is possible with biotechnology. Two examples of applying these technologies are the memorandum of agreement with the Amul Dairy Cooperative in India and the arrangements with institutions in the USSR and Mexico that were developed by the Monsanto and Eli Lilly companies.

Carliene Brenner in the August-September 1991 issue of the OECD *Observer* notes that it is "the developing world, where growing populations require an increase in both the volume and quality of agricultural production, that the pressure for technology innovations is most keenly felt," and further states that products based on plant technology are now expected to appear between 1995 and the turn of the century. The breakthroughs the author notes will "include plant diagnostics, microbio insecticides, tissue culture and micropropogation techniques, genetic mapping techniques in transgenic plants resistant to specific herbicides, viruses and insects."[1] Biotechnology cannot only improve yields and quality, it also can reduce the time it takes to develop new varieties, help monitor indigenous crop genetic resources, and identify genes with special properties.

Brenner acknowledges the progress that has been made in maize production in the selected countries of Brazil, Indonesia, Mexico, and Thailand, but asserts that biotechnological techniques alone are not enough for major breakthroughs to occur. Furthermore, local production rather than imports of seed has to be the dominant source of supply for most countries. For local production based on the new technology to occur, the development of a successful and strong private sector, of a technological infrastructure, of storage, distribution, and credit systems, of a market-oriented economy, and of appropriate protection of intellectual property rights must accompany the transfer of technology. Unlike most OECD countries, "none of the selected sample countries has yet adopted legislation on the broader issues of intellectual property rights (IPRs) for plants, animals and microorganisms which is an increasingly contentious issue in debates on new biotechnology."[1]

New forums are being created to aid in the transfer of technology, protecting national priorities, intellectual property rights, and the research and development investments of private and cooperative firms. For example, there is cooperation between private companies, local governments, and industrial research centers such as the International Center for Wheat and Maize Improvement (CIMMYT), the International Rice Institute (IRRI), and other centers that make up the Coordinating Group for International Agricultural Research (CGIAR), which has its offices at the World Bank. But there is tension between "farmer rights" to his or her own farm production and the contractual development of new variety production on those farms in cooperation with private firms. There also is an issue of ownership of germ plasm obtained in one country and used throughout the world. The question of the storing and maintenance of germ plasm banks for societal uses must be resolved. Finally, what is the appropriate technology for sustainable agriculture for forestry and economic development? This is the subject of a major conference to be held in Brazil in 1992 under the auspices of the United Nations and the International Business Council.

Biotechnology has the potential not only to become responsive to the nutritional needs of the developing world, but also to become a positive engine for enabling broad economic development to take place. Successful institutions such as the Amul Dairy Cooperative and the National Development Board of India that insulate themselves from political pressure can use both technology and aid without distorting existing markets. By including landless laborers and small- and medium-scale producers, they have enabled change to take place, and they provide an excellent model for both plant and animal technology transfer in the developing world. Superimposing biotechnology on this model should encourage appropriate and successful transfer of biotechnology to the developing world.

THE CASE FOR BIOTECHNOLOGY

In *The Case for Biotechnology*, Traverne points out that "the new biotechnology may well turn out to be the most dynamic of the great technological revolutions of the 20th century, surpassing even the development of computers and information technology."[2] He goes on to suggest that improving the quality of our life, our health, our ability to produce food and drink, and the environment are all possible with the new biotechnology. For example, with biotechnology, we can develop and produce transgenic mammals that produce human proteins in their milk. The purity and stability of these human proteins mean that factories for pharmaceuticals in animals and plants do in reality exist.[3]

As noted above, companies such as Pioneer Hi-Bred International and Calgene are now able to tailor-make crops suitable for manufacturers' and consumers' special needs in cereals, oil seeds, and vegetables. These products are tastier, more nutritious, and provide value-added qualities for which the manufacturer and consumer are willing to pay a premium.[4,5]

In addition to the development of crop and animal biotechnological successes, other breakthroughs have occurred in agriculture.[6] In 1990, fish farming sales, according to the Food & Agriculture Organization, amounted to over $22 billion. Dr. Thomas Chen of the University of Maryland's Center for Marine Biotechnology injected a cloned growth hormone gene into fish embryos, and almost half of them integrated the new genes into their DNA. The transgenic fish grew from 20 to 46% faster than ordinary specimens. Other researchers are trying to engineer disease immunity into their fish. The tests have the approval of the USDA with appropriate safeguards, for example, fences, dikes, screens, 24-hour surveillance, and poison to kill the fish if flooding occurs.

CONCERNS

(1) Unknown Dangers

Traverne[2] sets forth legitimate concerns about the new applications and uses of biotechnology. Some fear that new genetically engineered organisms might adversely affect the environment. He refers to Jeremy Rifkin's contention that scientists are playing "ecological roulette." Others are concerned that we may reduce the diversity of nature. Traverne replies that biotechnology may provide a more careful assessment of the effects on the ecosystem of introducing new animals or plants that might compete too successfully with native populations. He also points out that "the new techniques of genetic engineering are used to achieve a very specific result and produce a product designed for a very particular target."[2] Further, the fear that "small genetic modification might unwittingly turn a harmless organism into a dangerous pathogen is scientifically unsupportable" and biotechnology has promoted diversity as well as reduced it.[2]

(2) Small-Scale Farmer Pressure

There is concern that the new technology may not be "scale neutral" and may encourage large, sophisticated farmers to use the new technology first, thus widening the gap between the small and large farmers. Dr. Kurien and his 6-million-member Amul Dairy Cooperative have provided a model of how technology can be used by the landless laborer, the small-scale producer, and the medium-size producer without leading to their demise.

(3) Pressure To Use Technology

Some large-scale producers historically have been concerned by input suppliers' insensitivity to their hesitance to use new technology and have felt forced to accept it because of pressure by farmer competitors. This concern requires much more of a partnership between the input technology supplier and producer than has occurred in the past.

(4) Creation Of Surpluses And Depressed Prices In The Short Run

Widespread adoption of a new technology may add temporarily to a glut situation; therefore, appropriate safety nets for those pressured by short-term market adjustments may be needed.

Public Input In Evaluating Concerns

Because biotechnology can affect the whole food system from input supplier to ultimate consumer, appropriate controls and forums must include the public and consumer groups. The issues and public concern about them require the development of a credible neutral institution.

Case-By-Case Licensing For Each Planned Release

Concerning government review, Traverne notes that, "The Office of Technology Assessment of Congress in the United States has suggested different criteria of strictness embedding in different categories. A 'low review,' that is one with less stringent scrutiny should apply where, for instance, a new product is functionally identical with one already reviewed and approved, or is similar to a naturally occurring organism. A 'medium review' should apply where, for instance, a product is different in some ways but generally similar to previously existing products in general use. A 'high review' should apply where for instance, a product is a genuine novelty for which there is little or no previous experience. The Office of Technology Assessment also distinguishes between genetically engineered animals and plants for which it saw little risk and pathogens or pests used to distribute the genetically engineered products intended for environmental release, which they felt, should be more rigorously reviewed since they have the potential to affect the environment adversely."[2]

BIOTECHNOLOGY AND THE RESTRUCTURED GLOBAL FOOD SYSTEM OF THE 1990S

In the developed world the biotechnological revolution provides the technology that responds to changing global demographics and the development of market segments; in the developing world it provides the potential for commercialization of the food system.

The tailor-making of identity-preserved food products allows the input supplier to provide branded ingredients at the farm supply and farm level; these ingredients, in turn, enable both the branded food manufacturer and the private label food retailer as well as the fast-food restaurant to differentiate their final products to the consumer. Biotechnology allows food, feed, fuel, and pharmaceutical ingredient products to be produced at all levels of the value-added food chain. It also helps the food system to produce specialty products that help clean up the environment; for example, grain-based polymers can result in time-released packaging that is biodegradable either in sunlight or in underground waste disposals.

Biotechnology also can provide:

- Alternative sweetener, fat, and salt substitute products that are safe and can meet the nutritional needs of target groups of consumers;
- The opportunity for more flexibility in the value-added food chain so that fewer trapped resources remain underutilized, especially at the farm level;
- Substitute grain-based inks that allow safer recycling of paper products;
- Faster ways of regenerating forests and responding to other environmental needs;
- A broadening of the competitive structure of the food system by encouraging entrance of those firms that have not been involved in the traditional plant and animal genetic sciences.

Biotechnological participants in the value-added food chain should help forge a multidisciplinary, multifunctional forum that can improve the development of food and forestry policies on a national and international level and forge new links with existing institutions. Also, the biotechnological revolution should encourage the scientist to interact with those who will be applying that science in the market place.

Biotechnology is making a positive contribution to sustainable food production, as Dr. Robert Paarlberg[7] has noted in a recent paper prepared for the Sustainable Food and Forestry Project:

> It will be a challenge to show that such environmentalist anxieties about science-based farming are largely misplaced. The message must be sent that farming in rich countries has recently become more "sustainable" through science, technology and innovation—*not* simply by reverting to "tradition." Were it not for continuing scientific and technical innovation in rich countries, farming practices today would be just as destructive to soil resources as they were in the days of the Dust Bowl, and just as destructive of rural wildlife as they were when Rachel Carson published *Silent Spring*. Food production, meanwhile, would be much less abundant. There is an important lesson here for environmentalists and agriculturalists currently working in poor countries: unless urgent investments are made, in the people, the institutions, and the infrastructure needed to boost scientific and technical innovation in the farming

sector, both food production and rural resource protection will continue to falter.*

SUMMARY

The biotechnological revolution, along with the information, packaging, and distribution revolutions, have changed the global food system in the following manner:

Farm suppliers no longer supply individual inputs such as seed, fertilizer, chemicals, feed, equipment and credit. They have become innovators and transferers of technology, producing products that increase value, volume, and quality of the end products in a sustainable manner. Thereby, they improve production, processes, products, and distribution—while improving the environment.

All of this change takes place in a manner that is responsive to new consumer attitudes, in both developed and developing countries, with the private sector asked to take on a leadership role.

Commodity handlers have become food wholesalers of the world, and *food processors*, in turn, have become food, feed, fuel, and pharmaceutical ingredient suppliers. *Fast-food operators* and *food retailers* have taken a more active and anticipatory role in responding to consumers' changing needs; they have information systems that enable them to take a leadership role in the global food system and in the creation of national brands and premium private labels. These leadership roles include developing "environmentally and bodily friendly" products.

Structurally, the consolidation and globalization of the food system, and the complexity of the markets, processes, and technologies, have led to *new coordination tools*. These include horizontal and vertical partnering in order to be more competitive and responsive in segmented markets—for example, the Coca-Cola/Nestlé joint venture in beverages and the General Mills/Nestlé joint venture in cereals.

Finally, the changes in technology, structure, and coordination have occurred from the 1980s to the 1990s (see Exhibit 1). These changes call for a new type of global forum that cuts across disciplines and functions and that can provide a dialogue, enabling rules and regulations to be created in a credible manner. This forum must involve the scientific, medical, business, farming, consumer, and environmental communities, interacting with the appropriate domestic and international governmental bodies. The creation of such a forum could be one of the many positive outcomes of this important conference.

*Extended discussion on the role that scientific and technical innovation must play in rural environmental protection in poor countries is provided in the background documentation to the 1991 "FAO/Netherlands Conference on Agriculture and the Environment," Ls-Hertogenbosch, The Netherlands, 15–19 April, 1991.

Exhibit 1 The Changes in the Food System from the 1980s to the 1990s

1980s	1990s
Technology	
Primarily traditional plant and animal genetics with emphasis on improved yields.	Application of bioengineering to traditional plant and animal genetics. Entrance of pharmaceutical, chemical, and boutique crop and animal biotechnology companies as additional suppliers of technology to the global farm and food system in both the developed and developing food economies. Emphasis on improved yields, quality, flavor, nutrition, value, process, and safety.
Structure	
Farm inputs—one dimensional, supplied a unique seed, fertilizer, pesticide, insecticide, or animal health product.	Farm inputs—total technological companies serving the whole value-added system from producer, processor, and distributor to ultimate consumer.
Commodity processors and handlers served traditional grain, feed, flour, rice, oil seed, meat, poultry, dairy, and fruit and vegetable markets.	Commodity processors becoming unique food, feed, fuel, and pharmaceutical ingredient suppliers and global wholesalers.
Retail and fast-food operators specialized in particular types of customers and product lines and managed counter space and shelf space.	Consumer market leaders creating and developing new private label products and providing a complete consumer service and logistics system for market segments, in both the developed and developing countries.
Coordination	
Coordination of the system provided by traditional markets, government programs, future markets for specific commodities, vertical and contractual integration, joint ventures, and traditional farm, dealer, and wholesale distributors.	Coordination of the system provided by new horizontal and vertical relationships at the retail, food manufacturer, wholesaler, processor, farmer, and farm supply levels. Partnering, buying groups, and linkages into each others' information systems have provided just-in-time inventory management and immedi-

Exhibit 1 continued

1980s	1990s
Coordination, continued	
	ate response to individual consumer requirements. In the developing world, the private sector must create the coordination, grading, and pricing systems so that the developing world can commercialize its food system. The traditional markets coexist with the new partnering, providing price signals and risk management institutions as futures markets now exist in exchange rates, interest rates, as well as commodities. The distribution system is shrinking by the elimination of many subdistribution levels.
Regulation	
Regulation is handled by traditional agencies and departments such as food safety, agriculture, commodity exchanges, environmental agencies, and national research institutes.	Regulation is handled by all the traditional agencies of the 1980s, but new international bodies are being created as the European Community becomes more integrated in 1993. The development of biotechnology has increased the importance of scientific review boards. What is missing is a forum that cuts across disciplines and functions and that can provide a dialogue enabling rules and regulations to be created in a credible manner. This credibility demands the involvement of the scientific, medical, business, farming, consumer, and environmental communities as well as the appropriate domestic and international governmental bodies.

SOURCES

Carliene Brenner, Biotechnology in the Developing World—Lessons from Maize, OECD Observer, August/September 1991, pp. 9–12.

Dick Traverne, The Case for Biotechnology—Report by PRIMA Europe Commissioned by the Eli Lilly International Corporation.

Biotechnology NEWSWATCH, Volume 11, Number 17, September 2, 1991, page 1, "Goats, Sheep, Cattle carry genes for human proteins."

Remarks by Thomas N. Urban, Chairman and President, Pioneer Hi Breed International Inc. annual meeting, February 26, 1991.

Various news releases of Calgene, Inc., August and September 1991.

Mark Fischetti, "A Feast of Gene-Splicing Down on the Fish Farm," SCIENCE, Vol. 253 August 2, 1991, pp. 512 and 513.

Dr. Robert Paarlberg, "Agriculture and the Environment Food Production versus Resource Protection in Rich and Poor Countries," Harvard Center for International Affairs, September 23, 1991.

CHAPTER 1

The Paradox of Need vs. Overabundance

Jonathan I. Taylor

I, for one, see a basic paradox in the food system. On the one side there is a lot of well-founded data and evidence about food shortage, population growth, and a compounding problem which has us needing to double food production over the next 25 years to cope with demographic trends and nutritional needs. There is obviously some reality in that. But everywhere I go, whether in the United States, the EEC, Poland, or Hungary I see not food shortages, but in many cases food surpluses, and very often failures of distribution. We are all familiar with the surpluses: dairy products, oil seeds, feed, and food grains. In my own business we even have to cope with a 50,000-ton salmon mountain which severely inhibits one of our operations.

It is the same if you go to Poland. There is an enormous problem of disposing of the grain crops at the moment because Russia, the traditional market, no longer has the hard currency to pay. Somewhat the same situation applies in Hungary. In Russia you find that harvests may not have been as bad as was estimated, but the crops have not left the farm. Food grains are being used as feed, stored in the expectation of higher prices, or bartered for other commodities. So there is this paradox between the apparent need for more food and the possibly short-term problem of overabundance in many areas.

Robert Paarlberg is quoted in the background paper. I saw his recent testimony to the House Select Committee on Hunger. He brings an interesting, but rather traditional, approach to the world food problem, one with which I sympathize. It is that solutions are not going to come from revolutionary new technology, but from doing better what we are already doing, or even from just continuing the things that we are doing at the moment. He puts particular emphasis on building rural roads and repairing those that have been built; maintaining irrigation systems; constructing post-harvest storage; planting trees; maintaining research programs and extension programs; and, in the public sector, maintaining the World Bank and the AID support for the whole edifice of international agricultural research institutions that have done so much for agriculture.

In the last few years, public institutions such as the World Bank and AID have believed that structural adjustment, free markets, and letting prices rip can solve any problem. This is true in certain situations. But if there are no rural roads, if there are no storage facilities, if the irrigation system has fallen to pieces, free markets are not going to do much for the farmer. I would emphasize the need to maintain what is already there and to find the investment that it requires.

Regarding the unpredictability of future needs, I shall share with you a recent snapshot. I was recently at a Booker board meeting, and we were confronted by some figures that I found mind-boggling. We have projects in Zambia and Zimbabwe, and two or three months ago we tested the entire work force for HIV. Between 30 and 50% of the work force came out HIV positive. This is a population catastrophe which started in Uganda, Rwanda, and Northern Tanzania, and is moving south very rapidly. It has also moved into West Africa with a slightly different form of HIV. The demographic implications are unperceived at the moment: maybe in certain parts of the world we are not looking at a population explosion, but a population implosion. This is one of the world's uncertainties that perhaps has to modify the way we see the need for technology.

There is an issue which I am sure we are going to confront over and over again and it is summarized in Eurospeak as "the fourth hurdle." To bring the issue closer to home, I would suggest that sweeteners may have to confront the fourth hurdle. Let us take high-intensity nutritive sweeteners. I think that thus far NutraSweet® has found an incremental market, and it has serviced this new diet-related market. It is welcome and plays an important role.

But suppose that someone were to develop a sweetener that had a very high intensity, at a cost substantially lower than conventional sweeteners such as sucrose, fructose, and glucose. Suppose it had the appropriate flavor and texture characteristics, was cookable, was in crystalline form, and could replace conventional sweeteners both for domestic and industrial uses. How would we feel about it? How would we reconcile it with the future of cane and beet farmers, corn growers, and sugar workers in developed and developing countries? I have cited sugar because it is close to home, but the point applies equally to coffee, cocoa, vanilla, gum arabic, natural dyes, many of the spices, and other high-value products which are produced by farmers throughout the world.

Technology may produce a totally altered pattern of comparative advantage. One instinctively feels that we should not stand in the way of technical progress. But there are enormous social and economic implications in such progress, and this is what we call the fourth hurdle. We may have products that are safe, have high quality, and are effective, but do not pass the fourth hurdle, that of need. I wonder what we all feel about this dilemma. I have no answer. I acknowledge the enormous benefits and progress that many branches of biotechnology have brought to the food industry, but I see them as supplementary, as means to an end, not as ends in themselves.

CHAPTER 2

The Industrialization of Agriculture

Thomas N. Urban

It is useful to view your business and your industry from a distance. I see agriculture going through an industrialization process similar to that which manufacturing went through at the turn of the century. I have been asked to identify and to discuss technologies which are at the heart of the industrialization process. I am going to touch on three technological developments: classical genetics, with which we are intimately involved; molecular biology, with which we are also all familiar; and the information technology revolution that certainly is upon us.

It is clear that technologies are only useful if they increase productivity. So it might be useful for us to go back and ask ourselves where the major productivity gains have come from in production agriculture. The University of Minnesota, which has done a significant amount of work in the area, estimates that 50% of the productivity gains in agriculture have come from a group of inputs: the use of fertilizer in 1940–1960s; the chemicals with which we are all familiar; timeliness, which we all know something about; and, of course, general management ability. It turns out, however, that the other 50% has come from genetics. Genetics, in fact, has contributed at least 50% of the productivity gains that we have found in production agriculture over the past 50 years.

When I go out and talk to farmers and I ask them where they think they are going to achieve productivity gains in the future, they sit back and think, and they are somewhat perplexed. They look at the opportunities in fertilizer and do not see much. They see ways to use less of it and apply it more efficiently, but they do not see big advances in that area. Chemicals have made an enormous contribution to agriculture, but farmers do not foresee significant productivity gains from chemicals as they look at their weed and pest problems. In general, farmers are concerned with timeliness, and they do a pretty good job of being timely. However, they do not see major gains in timeliness in the future.

They do see themselves becoming more efficient managers. But when they think about it, and we agree with them, many of the productivity gains are going to come from genetics. Genetics are going to make a dramatic difference over the next 25–30 years in agricultural productivity. That means, of course, that farmers are going to spend more money for genetics than they have traditionally spent, because variable costs tend to flow toward productivity gains. They will be spending relatively less on other inputs.

Classical genetics will drive most of the genetic gains in the near term, in my opinion, not molecular biology. Whether one talks about health, environmental concerns, taste, or production concerns, classical genetics will continue to make the significant contribution. We are changing the fat profiles of foods, for instance, in an attempt to respond to consumer needs. We are doing that with classical genetics, not molecular biology. Taste concerns are being met with classical genetics; food corn is a good example. In the production area, starches for the paper industry and oils for manufacturing are also being improved with classical genetics.

We are limiting biotechnology to those few areas in my opinion that classical genetics cannot attack productively. Lepidoptera resistance, in cotton for instance, is a significant improvement that will reduce chemical use. Viruses in alfalfa are an example. Corn borer and rootworm in corn are specific targets.

But after listing these opportunities, one's voice drops quickly as one looks for profitable opportunities using molecular biology. Many of the opportunities that exist today reduce production costs. That is, they substitute genetics for chemicals. On the animal health side, one can clearly make some progress. Trying to put methionine in soybeans to enhance feed quality is an example. Again, one is substituting a naturally occurring product for artificially added methionine.

One of the greatest contributions from biotechnology, however, is the enhancement of the process of classical genetics, RFLP technology, which allows us to perform classical genetics faster and more effectively. It may very well be that our knowledge of molecular genetics will make as much or more contribution in the exercise of classical genetics than it will in the more romantic area of gene transformation.

Information technology is driving information from the consumer back through the producer distribution system to the farm. That link, that bridge of information, is becoming stronger and stronger. We are getting to the point where we can almost identify certain farmers with certain shelf space. In other words, through contracting, the movement of capital, farmers are becoming more and more attached to a particular consumer outcome. We have seen this in the broader industry; we are seeing it in the cattle industry; we are seeing it in hogs; and we are going to begin to see it in feed grains.

It turns out, then, that farmers are going to be a more integral part of consumer demand. As opposed to producing a commodity that is then transformed by some portion of the food chain, farmers themselves are going to receive a premium, or at least a contract, for specific characteristics that go around the commodity system directly to the processor or the supermarket.

Like many of you, my company does business in many countries. We are finding that these three technologies—classical genetics, molecular biology, and information technology—are, in fact, moving aggressively across the world, both in developed and developing countries. From our perspective, we are all part of the same system. The flow of technology, capital, and information up and down the system, from the consumer back to the farmer in developed and developing countries, is a reality. We as a company are trying to respond to that flow by correctly positioning our people and our technology dollars.

CHAPTER 3

The Biotech Paradox

Roger H. Salquist

The topic that was described for this conference was how "revolutionary" new technologies will affect the structure and use of the global food system. To a biotechnologist that statement alone creates what I call the biotech paradox. I spend most of my time telling Wall Street how revolutionary this technology is and how it is doing all of these dramatic new things. Yet, from this point forward I am going to spend most of my time trying to tell consumers that it is not revolutionary at all; it is really rather mundane, evolutionary, hardly any different from classical genetics. That creates one heck of a problem when biotechnology is the total heart and soul of your business. I do not know that the ag-biotech industry has totally come to grips with that paradox yet. I would say that we have erred, if anything, on the side of evolutionary rather than revolutionary if you look at the values of ag-biotech stocks relative to biomedical stocks.

What I would like to do is compare my vision for the future of this technology five years ago with my prognosis for the technology today. My prognosis in the last 12 months has plummeted, and I have serious doubts, at least in North America, that recombinant DNA technology is going to make much of an impact in the near future for some fundamental structural reasons that I shall describe.

On the positive side, though, anybody who has followed this field would have to conclude that what the technology has accomplished and is delivering today is far more than the most optimistic people would have predicted three or four years ago. We now have had field trials in the United States in over 30 states with more than 15 different crops, a clear demonstration that you can use this technology to do a considerably broader range of things than Tom Urban suggested it would do. There is no question that any biotechnologist in this field relies on classical genetics to take the lead in developing plants that are hardier, more fit, and better on basic productivity issues. But I think that the agenda of things that can be accomplished with recombinant DNA is, in fact, considerably wider than Tom Urban suggested.

It is ironic that a lot of the attacks on this technology are coming from the environmentalists, when in fact the new biotechnology is the only tool we have to significantly enhance future progress in, for example, integrated pest management and no-till farming. The only way to achieve broad increases in no-till farming is through the use of herbicide-tolerant crops with environmentally compatible chemicals that allow you to avoid pre-emergent herbicides and use post-emergent herbicides. It is a tool that one does not get with classical breeding, and yet it is not something that people typically recognize.

The power of the technology continues to amaze me. There is nothing that we undertook five years ago, including our most technologically risky projects, that we have not been able to do. For example, our most complicated task was to clone the genes in the pathway that causes plants to synthesize oils. We have now cloned six genes in that pathway. We have plants that express those genes. In fact, we have our first five field trials underway with genetically engineered rapeseed that accumulates very high levels of stearate, a fatty acid that those plants typically would not produce without recombinant DNA. These things suggest that the technology is real.

So what are the problems? To put it into perspective, Calgene is now 10 years old and we are still two years away from our first product using genetic engineering. We have raised $110 million in equity financing and $45 million in corporate funding. We have invested about $80 million in research and development over the last eight years and have a current research and development budget totally focused on recombinant DNA of about $12 million per year. That consists of about $7 million for research and $5 million for development.

In terms of measuring progress, we have now completed three years of field trials with our most advanced constructs in both tomatoes and cotton. We are the first and the only company that has a petition pending before the Food and Drug Administration for a genetically engineered whole food. As you know, the Food and Drug Administration has approved one genetically engineered food product to date, chymosin, a genetically altered enzyme for making cheese. But they have not heretofore addressed the issue of a genetically engineered whole food. I might also note that in 1991, under one permit, we did 34 field trials in 12 states with cotton.

From a structural perspective, one of our fundamental strategies has been corporate alliances. It was quite clear to us as a start-up company in a business as complicated as agro-business that we could not do it alone. We have developed, using our technology, a series of corporate alliances. The most interesting is our alliance with Proctor & Gamble Company: we are doing research for them, but our agreements provide that we shall be the producers of product for them and supply that product downstream. Not the least of these opportunities will be as suppliers of an alternative basic ingredient for the detergent business, of which they use 400 million pounds a year. When we start producing that in U.S. agriculture, it will make a structural difference in the places we choose for production.

We have another alliance in the oils sector with the Ferruzzi Group of Italy, which includes Central Soya. We also have our partner in fresh produce, the Campbell's Soup Company. Campbell's Soup continues to have the rights and the obligation to commercialize our technology for processed tomato, while we shall do it on our own in the fresh tomato market. In the potato sector we have a strategic corporate partnership with the Kirin Brewing Company.

I believe these are new types of relationships for the food industry. In the pharmaceutical industry they have been quite common over the last 10 or 15 years. Developing technology in structures like this in the food industry is a significant new trend.

I want to return to why I have become negative. To be honest with you, I think that what happened between 1985 and 1990 was tremendously positive for the acceptance of this technology, both at a government level and in the public sector. Since the publication and adoption in 1986 of what is called the Coordinated Framework, of which my colleague John Cohrssen was a principal architect, we have been able to take this technology into the field without a single piece of legislation, without a single incident, and with the wheels of government working reasonably effectively. We have also, during that period, had a validation of the safety of the technology through a major study by the National Academy of Sciences called "Field Testing of Genetically Modified Organisms," which was published in 1989. It concluded that the existing structure was adequate to monitor and regulate this new technology, and there was no inherent difference between this technology and classical breeding to develop similar traits.

During that same period of time we had the successful start-up and public financing of seven or eight different biotech companies. I do not know what the amount of capital raised was, but it was on the order of $300–$400 million. This had never been done before with start-up companies in this field.

By the same token, the hundred or so field trials that I mentioned earlier took place without a single protest. These trials involved plants. Opponents like to think of the frost ban several years ago as an example of public dissatisfaction and distrust. However, that was a blip, not a trend. Everybody forgot about it. In fact, it is one of the nails in the anti-biotech coffin because every dire prediction that environmentalists made at that time had no basis. They were simply false, and none came true. One of the benefits of this is that the knowledgeable national media now give far less credence to those kinds of comments than they used to.

Thus, things went along pretty well through 1990. A bill introduced in Congress last year by Congressman Roe that would have created a whole new network of counterproductive regulations died. Mr. Roe moved to a different committee and nobody picked up that particular piece of legislation. I thought that was quite positive.

This era of progressive development was capped with the publication in February 1990 of the President's Council on Competitiveness Report on National Biotechnology Policy. This was a very positive statement that said we need to minimize government regulation and maximize the introduction of this technology. But, frankly, that is about as far as it went. What has happened since then is that the Council on Competitiveness, which is supposed to be the government's lead agency in this area, has been leaderless. In some 10 months we have gone through three different heads of the Biotechnology Working Group at the Council of Competitiveness, and there has not been a single spokesman within the federal administration who has stood up and said, "I am in charge; this technology is good; and we are going to assure that it gets to the market." This is creating an ideal void for legislation to be introduced and relates directly to Roger Porter's position about the need for uniformity (see Introduction). The apparent confusion at the federal level fosters state and local legislation.

During this same period of 18 months, we have had what I would characterize as ideological wrangling over how the government is going to do what it is already doing. The focus of that debate is what is called the Scope Document. Progress in moving forward to final commercialization has been completely log-jammed first at the EPA and at the USDA, and it will be at the FDA if we cannot get our act together and acknowledge that in principle you regulate processes, not just products. The fact of the matter is that the public, at least in the United States, needs to have the federal stamp of approval on this new technology no matter what the product. That is only going to happen if you acknowledge that fact and call it regulation.

The first products of this technology are going to have to be reviewed. You can call that regulation or oversight if you want, but it has not been happening. We really have to make something happen in Washington, and that is my challenge to John Cohrssen.

In my opinion the other negative change that has happened in this period is in the leadership at the FDA. The stated objective of David Kessler was to enhance public confidence in the FDA and the safety of the food supply in this country. In fact, his actions have had the opposite effect. I think he has undermined the credibility of industry by the capricious nature in which he has undertaken his enforcement program, and I think that it is going in the wrong direction.

Dr. Kessler recently spoke at a biotech meeting and was asked, "One of the principal problems you cite is the lack of resources at FDA to get the new technologies to market, yet your first act when you got new resources was to hire 200 new detectives to go out and plow through people's storage areas and supermarkets. How do you rationalize that?" He did not have an effective answer.

The final point regarding biotech is the external measure of this technology, the stock market. The technology has proven itself; yet there are no new companies coming along, and there are no new financings emerging to foster this technology. The first six months of 1991 was the biggest biotech financing window in history: more than $2 billion in capital was raised; $550 million of that was initial public offerings for new companies, not one of which was in agribusiness despite all the scientific progress. There was a billion and a half raised for existing biotech companies, and less than 5% of that was raised by agribusiness or ag-biotech companies. That is a tragedy given the proven power of the technology and the potential opportunities that exist in this industry.

CHAPTER 4

The Role of Government in Technological Change

John J. Cohrssen

Ray Goldberg's chapter discussed the pace and magnitude of change. Indeed, 20 years ago recombinant DNA was just a glimmer in Cohen's and Boyer's eyes. Today biotechnology is a $2 billion industry and that is just the beginning.

During the past 20 years the government's role in biotechnology has also grown exponentially, causing the Vice President and the Council on Competitiveness to examine it. The Council was established by the President to continue U.S. economic progress by maintaining and improving international competitiveness. One goal was to ensure that the government gets out of the way so that the private free market can work. It was chaired by the Vice President, and other members included the Attorney General, the Secretaries of Treasury and Commerce, the Chairman of the Council of Economic Advisors, the Director of the Office of Management and Budget, and the President's Chief of Staff. Other cabinet members and senior officials were invited to participate in particular meetings.

The Council established a biotech working group in 1990 that started looking at the landscape and climate of the government's role in biotechnology. That group issued a report that has been widely read overseas. If imitation is a form of flattery, I have seen a Japanese version that has the same four-color graphics as the English version, but with a Japanese section attached to the back as well. The report states that in biotechnology, as in other technologies, the role of government is to deal with areas that are sometimes called market failures. It is to provide needed support in limited circumstances to activities that offer large potential benefits to the economy as a whole, but do not offer the prospect of adequate profit to any particular firm that might undertake it.

Another role of government is removing unnecessary and artificial barriers to profit market functioning, including legal and regulatory barriers. In biotechnology the government has played a substantial role in strengthening science and the technology base, establishing risk-based scientifically sound regulation and maintaining a flourishing free market with available capital and protections for intellectual property.

The science base continues to grow. Currently the Office of Science and Technology Policy is doing a cross-cut look at the range of government funding. About $3.6 billion is earmarked for biotech in fiscal 1991. I cannot tell you what will happen next year, but I suspect that funding is not going to go down. United States government-funded research is broadly available to the private sector through a variety of technology transfer activities.

Concerning regulation, Roger Porter (see Introduction) mentioned the four principles, which are that regulation should be risk-based, science-based, and product-based, but not process-based. In Washington we also hoped that it would not be turf-based, because regulation itself can become an industry, with vying government agencies seeking to have a substantial role over sister agencies.

We are currently seeking to complete the government's overview of the regulation of the biotechnology industry. I share Porter's frustration that it has not moved more quickly. I have been involved with this for a number of years, and it takes a while for things to move along. One of the concerns is that with technology advancing so rapidly, we must avoid developing regulatory regimes that cannot keep up with the technology, that would be regressive and locked in a particular channel. This is difficult, because we must also provide the regulatory certainty that is needed. We do not want regulation to be a drag, to favor the older, less efficient technologies.

One of the challenging issues throughout is to define what is new. What is new about a food product? Is it new to the extent that it should be regulated? Roger Salquist pointed out the schizophrenia of going to the capital markets and describing something as new and exotic, while telling the regulators that this product is business as usual. Maintaining one's sanity while doing that is very challenging.

The Council is currently producing a series of road maps for the commercialization of technology, and these are taking longer than we had expected. We shall have a road map focusing on plants, one on microbes, and one on animals. Moreover the United States is working with the other 24 industrial nations in the OECD to try to achieve risk-based principles for our industrialized partners. The OECD countries seem to be going in the same direction as we are in our conceptualization. Certain European countries and even the EC have this concept of the "fourth hurdle." There are some countries that have tried to regulate process, not product, and then found that it was very hard to describe the process.

As we move ahead, we see a limited role for government, both domestically and internationally. Ray Goldberg spoke of credible neutrality. I dare say the government is not always credibly neutral. As we look forward to either greater acceptance of this technology or, maybe more importantly, less rejection of it, the question is how to achieve credible neutrality. How can the private sector or other parts of the country deal with establishing this type of credible neutrality?

CHAPTER 5

The Commercialization of Biotechnology and the Flow of New Knowledge

Jozef S. Schell

Let me tell you about the subjectivities that will unavoidably bias the comments I make. I am a European, and I think of myself as a "convinced" European. I am a scientist, and my area is the molecular biology of plants. I am very much convinced that the responsibility of science is to contribute to the elaboration of new technologies, particularly technologies that have the potential to contribute to the solution of major problems.

In the area of plant biotechnology the progress that has been made thus far justifies the most wildly optimistic visions of what science and technology can do for a world that rapidly will be confronted with unsustainable agriculture and global problems of overpopulation, even if population predictions may be off because they have not taken account of unexpected events such as major epidemic diseases.

I think we are under great time pressure to make progress. I am convinced that we do not have all that much time to make critical breakthroughs. We urgently have to resolve the tension between productivity in agriculture and sustainability of natural resources. In many cases this means that we need a productive agriculture that will not lead to the depletion of natural resources or the destruction of a desirable environment. I therefore agree with what has been said before. Improvement of germ plasm by breeding, including molecular breeding, is one of the effective ways to achieve the previously mentioned goals. This is where science has delivered and is continuing to deliver. I have little doubt of our capacity to continue to provide the knowledge, the tools, and the human resources to achieve the goals that have been discussed and written about.

Keep in mind that any property of plants, be it growth, yield, or response to environmental stresses, is continuously determined by the genes carried by a particular crop. We now understand how genes work and that we can use genes to obtain predictable results. We have found that in order to modify properties of living organisms, we are not limited to the use of genes that actually evolved in these organisms. We can, in fact, use any genetic property, whatever its origin, and introduce this as a new trait in plants. Most of the examples that you will hear about in talks or read about in books and specialized journals, illustrating the success of molecular plant breeding, have been achieved by making bacterial genes work in plants. These examples should help us understand the extraordinary scope that is now open for plant breeding.

Given enough time and resources, any of the goals that breeders set themselves, such as those already discussed (improved plant protection, stress tolerance, increased nutritional value, etc.) can be achieved. The bottleneck is no longer scientific knowledge and technology, but all the steps needed to bring a new product to the market. These problems are, of course, different, depending on whether we talk about the industrialized or developing world. Unless the transition to commercialization and practical application is achieved soon, however, there is a considerable danger that the flow of new knowledge will stall for lack of support from the public and subsequently from research organizations.

My perception of this problem may be somewhat biased because I happen to work in Germany. In terms of society questioning the value of science, new technologies in general, and biotechnology in particular, what we are going through is pretty dramatic. The responsibilities are manifold, but a large proportion of responsibility must be allocated to industry and to government institutions, particularly those involved in regulatory processes. Those who have the responsibility to make new knowledge actually serve society should know that society at this time no longer automatically equates scientific advance with progress. Yet this is the premise by which most scientists work. Society now wants to know, in advance as it were, the consequences of the application of new technologies.

Therefore, if industry does not do well with a new technology, society is likely to discontinue its interest and support for the scientific research connected with this technology. My major concern is that we are presently in some danger of weakening a science base that is very important to our culture and to our economic survival.

Let us look at the example of a newly bred crop, the use of which would contribute to a reduction of the negative environmental impact of high input agriculture. I am talking about the use of crops expressing the BT insecticidal toxin to provide protection against insect attack. One of the first commercial products planned to be ready for the market in 1995 is "BT-Cotton." It will be important to introduce this product wisely. If, for commercial expediency, BT crops are widely disseminated without the proper agronomical, regulatory, or pest management methods, we risk the emergence of resistant insect pests. This would not only prevent the further use of this important plant protection system, but could produce a backlash that might discredit the use of molecular plant breeding in a more general way.

This would be all the more a pity because proper pest management methods are readily available that could prevent such a scenario. For example, one could combine several different insecticidal principles in the same crop, or one could combine the use of BT crops with a wise and moderate use of selected insecticides. It is not scientific knowledge and research that would be responsible for a breakdown of resistance if proper pest management is not applied. It would be a lack of regulation, or the exaggerated and narrow-minded pursuit of short-term commercial interests. Yet all of this would also reflect badly on the science and the technology as a whole.

I know that the industry is well aware of these considerations. Yet I want to stress the fact that science and technology and their transfer to the marketplace are more intimately linked than ever before. This link must be duly considered by the scientific community. One should not see industry only as a potential source of funding, especially when other sources of funding are drying up, but also as a partner with whom one shares responsibility for the future of the whole enterprise. And if industry wants to make sure of access to ever more advanced basic knowledge on which to base its technology for the future, it should use these technologies in ways that are responsive not only to short-term market imperatives but also to long-term socioeconomic consequences.

I would like to address another aspect that has not yet been discussed here. We all know that nothing will succeed in this world if the rules of the marketplace are not satisfied. But it would be a mistake to think that by simply following consumer and market-oriented principles, one will actually solve the urgent global problems faced by our world. Here I see a role for an increased and more efficient and responsible interaction between industry and responsible international regulatory agencies.

There is no substitute for efforts by the science community, as well as by technology-driven industries, to take the dialogue with society seriously, and to reflect on what it is that we jointly want to achieve for the future. Both science and industry must ultimately show society that these scientific-technological goals of ours are needed and justified and that we can be trusted to proceed!

DISCUSSION OF PART I:

PROFESSOR GOLDBERG: Everyone talks in glowing terms about the future and how we have joint responsibility and how the structure is evolving, but we really have not spent much time on the bottlenecks and the problems we are having. Perhaps some of the questions will address these issues as well.

MR. TEASLEY: I think one of the big issues that we live with is a world where innovation and ideas get attacked and often not with good science. The public relations and political arenas get deeply involved. Industry and science need a greater database to support what we are doing. How do you get innovation out into a world where there is a lot of attack with bad science?

MR. URBAN: I think it is sometimes important to have perspective. The Homo sapien has been innovating over, I suppose, the last 3000 years. I can imagine the Inca Indian youth arriving at the temple bringing a new form of corn that he had discovered (or maybe his crossing activities had improved it) and being denied the right to plant in any of the tribal gardens because it would clearly destroy the traditional way of farming. I can think of many innovations that we have opposed because we are conservative by nature. You know about the Luddites as well as I do. The combustion engine is a further example. You will recall in the 1920s there were 25 million acres of oats and a huge industry built around the horse. All of that has disappeared, to be replaced by engines and soybeans.

So I am not really worried about how this process goes forward. It will go forward. If the technology is useful, it will be accepted; and I think it is simply a question of working at it on a day-to-day basis.

The only difference is that today it may happen a little faster than it happened in previous generations, simply because our communications are better than they used to be. Is that speed of change in itself a concern? I suppose it is; but, frankly, I hate to sound unexcited or uninterested in this problem. I think it is always going to happen if the technologies are, in fact, useful. That simply is how the Homo sapien has functioned for some 3000 years.

MR. SALQUIST: I have to say that I am an optimist about what I would call the increasing public sophistication in dealing with complex technologies. The phrases used in the lay press over the last five to six years dealing with this technology are increasingly sophisticated. Read *USA Today* and you read articles about metastatic tumors, chromosomes, genes, monoclonal antibody, and so on. They are coming to be common parts of the language. In two to three years every person you know is going to have had their life affected by biotechnology—either having a monoclonal blood test, or having EPO for kidney dialysis, or having a friend who has had a heart attack cured by TPA or similar new biomedical products.

My strategy as a company all along has been to let the medical area pave the way; let people see the benefits of this technology and become familiar with the nomenclature. It is going to become somewhat routine. From a product standpoint, we would hope that our tomato is so boring by the time that it comes to the market that nobody cares. The only question they ask is, does it do something for me? Does it have a benefit? We are doing focus groups right now with people on genetically engineered fresh produce. I would say 90% of the response is, "I do not care how they got there, but if it tastes better and lasts longer, I want to buy it." That is highly encouraging to me.

The other aspect of public education that I think is incredibly important, in which we have made spectacular progress in the last five or six years, is education of the media. If the media has not been sensitized to what the issues are, and what is good science, and what is sensationalism and pseudoscience, you have a big problem. In the United States today we, in fact, have quite a good group of national writers in the most influential press who are both educated biotechnologists, sensitive environmentalists, and good consumers. People like Donna Walters at the *Los Angeles Times* and Malcolm Gladwell at the *Washington Post*. There is a fairly long list of such press people who have become sophisticated. So at least today, when this technology is attacked, we get a balanced story. That is major progress.

PROFESSOR SCHELL: I nevertheless think that we are facing a crisis and that time is running out on us. Therefore, if one does not use the technology at its optimal potential now, the global problems that one shall be confronted with in the next decade will be of such a magnitude that we may not be able to deal with them at all. The time pressure under which all of this has to happen is greater than ever before. This is why I do not feel comfortable with the idea of "let's simply muddle through." I think it is the responsibility of each generation not to muddle, but to do its utmost to show its sense of responsibility for the future.

From my experience in Germany I think that the situation is beginning to improve. Germany is moving from a situation and mentality in which the luxury of a no-risk environment and a no-risk society was seriously considered, and indeed politically requested, to a more realistic attitude. This vision of a no-risk society was, in my opinion, a short-lived historic aberration. I expect and hope that attitudes will change rapidly.

I also want to make a scientific comment. Contrary to what has been said, both goals and hazards are remarkably predictable with these technologies. Those of us involved in the development and application of these new technologies should as an integrated team claim responsibility and convince society that we are best qualified to be given the responsibility for their use.

MR. SHAPIRO: I have a general sense of concern and perhaps less optimism that society is going to accept scientists or government as arbiters of a set of issues that concern them. It may very well be that there is, in fact, no better way than to have rational risk assessments and benefit-risk analyses performed by people who are competent to do those, and who operate with a sense of conscience and responsibility. But it is not at all clear to me that Western society, at least, is prepared to recognize either the moral or technical ability and right of any elite to make those decisions.

MS. CULLITON: I think, first, that the attacks that we have seen in the past decade on science are not necessarily rooted in an antiscience stance per se. Rather, they are a social-political attack on the establishment, which we have seen in lots of areas of society. Science just happened to come in for scrutiny after there were attacks on other issues—civil rights and liberties, for instance. It is all part of the same cultural mindset. Some of that is beginning to change, and I am not at all persuaded that either the press or the general public is actually against science.

The other thing that is important and that the panel alluded to, with respect to the general public's acceptance of all of the new technologies, is that once some company has produced a product that the public at large really wants, the opposition is likely to fade very, very rapidly. Most people are not interested in the subtleties of some of the issues that are before us. But if you really do produce a tomato that somebody would want to eat, that would make an enormous difference in public response. People care about tomatoes; they do not care about germ plasm. Similarly, in the medical arena, once gene therapy reaches the point that it is actually making sick people well, all of these abstract arguments we have about the morality of it and the control of it will be subsumed in acceptance of the fact that technology is now doing something useful.

At the moment we are in a transition. The press is no longer following the liberal agenda in quite as much of a knee-jerk fashion as I think it used to. Technology is also in a period of transition. Once it moves out of that transition and starts producing products that people can identify, that will change the public perception more than anything else.

MR. BLOBAUM: It seems to me that a lot of the anxiety is about the regulatory process and not about the technology itself, and that is a very important distinction. The ability of agencies to provide a real safe food guarantee is something that the public is not convinced is in place at the present time. Consumers are not convinced that all of the decisions, such as those in the pesticide area, are based on good science. I think scientists would be hard put in many instances to make that case.

From our perspective, we would say that we need to give more resources to the whole regulatory function, and I think industry would gain as well, by having a system that the public trusted. Some of these side attacks would no longer be made if the system really worked.

DR. MILLER: I would like to address primarily the issues of regulatory barriers to acceptance of this new technology. I think there is a tendency for us to look at the world as it was, not as it is. We look back at the problems; we look back at the attitudes of regulatory officials. We are not really looking at the fact that these, like everything else, tend to evolve as information is gained and confidence grows among regulators.

I think it is unfair to say that regulators are looking at genetic modification as a process rather than as individual products. Recently a joint WHO-FAO committee on the issue of regulation and safety of biotechnology-produced foods established as its first principle that the process by which the product was produced was not of immediate relevance to the issues of safety. It was the product that was of concern to regulators. This joint committee of experts was put together the way it was precisely to assure senior governmental scientific input, and consisted largely of regulatory scientists from the United States, Canada, the United Kingdom, and several other countries. The conclusions of the committee were the result of a unanimous decision.

On the other hand, the committee also pointed out another generalization. If the product is the endpoint and the process is irrelevant, then the same standards of safety ought to apply to those products produced by conventional breeding or any other process. There is no particular reason to assume that a product produced by conventional breeding is inherently any safer than a product produced by genetic modification. Both have to be looked at from the point of view of experience and knowledge and the information that is available.

There is another point I want to make about the attacks on the scientific community. Part of the problem is that we are confusing safety and regulation. These are two separate issues. Safety is an area that is amenable to at least relatively objective scientific evaluation. Regulation is a process that incorporates the issues of safety, but also superimposes a whole series of other issues that, until recently, we have tended not to speak about when considering regulatory policy. We always speak about safety issues, but they are not alone; there are economic issues, social issues, and so on.

The problem we are dealing with in part is an attack on the establishment. I tend to see these attacks, in part at least, as a kind of a neo-Luddite antipathy to new technology. Remember, it is not only biotechnology that has been attacked, but any new technology. Food irradiation, for example, an extremely powerful tool to assure safe and quality food, has been savaged. It is virtually unused in the United States and is not really widely used around the world, excepting certain areas, because of these unfounded fears. The attack is on the new technology, on the fact that the establishment has said something is good or something is not so good and has to be replaced.

The bottom line is that somewhere along the way there have been changes, and we are recognizing these changes. Regulatory officials are recognizing that biotechnology must be evaluated on a case-by-case basis. But I also agree with Roger Blobaum that you will get what you pay for in your regulatory officials. If they are not competent scientists, confident in their ability to make judgments and defend them, they will say no, whether you are talking about a process or a product.

It is extremely important that if this technology or any new technology is going to be utilized, irrational barriers be removed, and that the Administration, Congress, and the food industry come to grips with the fact that a strong competent regulatory agency is the only way to go. That simply takes money.

DR. ARNTZEN: Let me make a comment first about the technology. Most of what we have done in crop improvement in the last decade has focused on alternatives to inputs: virus resistance, herbicide resistance, insect resistance, and so on. In most of the university laboratories and many industrial laboratories we are now also focusing on post-harvest traits, quality determinants, things that will make differentiated commodities. Certainly, as Pioneer knows, within 10 years we will not just be dealing with No. 2 Yellow Dent Corn. We will be producing corn that is targeted with nutritional enhancement for the swine industry or the poultry industry, or special processing traits for the high fructose market, or for industrial raw materials.

The technology component has shifted, and it is this next wave that is going to be much more important in genetic engineering. The new technologies are going to make possible a number of changes that perhaps will be of less interest to the government regulators (enhancing the nutritional content of chicken feed, for instance) than it would be if the modified products were being fed directly to humans.

I think the best way we can be concerned about the regulatory and public acceptance issues is to make sure that all components of the "food chain"—from producers to grocers to the consumer—are involved.

At this particular meeting we have good representation of the input side of agriculture. We have the seed business with Pioneer and chemicals with Monsanto and others. We have good representation of the food industry. But I did not see anyone here from the American Wheat Producers or the Cotton Council or the Cottonseed Crushers or the Farm Bureau. There is a gap in here, between the guys who sell to the farmers and the people who buy from the farmers. There is limited representation by the farmers.

I think if the farmers do not get involved with this and the organizations that I spend most of my day interacting with—the Texas Cottonseed Crushers, the American Cattlemen's Association, and others—if they do not feel they are a part of this process and benefiting from it, if they do not feel that their livelihood is going to be enhanced, they are not going to be strong supporters. There have been a few farm organizations like the American Farm Bureau—not necessarily state organizations—that have been very supportive of biotechnology. Maybe we will blame it a bit on some of the organizers of this meeting that they did not include more representation from the producer segment of the food and fiber industry.

As we are dealing with public concerns about what is going on here, it is very important that the farmers see that within their lifetime—and they are primarily concerned about the next five years—that there is going to be something that comes out of all of this that is going to benefit them. The first wave of input replacement technologies is going to help reduce the cost to their production systems. But farmers are really more interested in the next wave of technology. If you can get differentiated commodities that will demand a higher price in the marketing system, especially, in the case of the United States, for targeting export commodities, I think they are going to be much more excited. Then we will get much more support influencing government regulators and others.

MR. COHRSSEN: I want to mention reaction to the concept of the more responsible press. I agree the press has become more responsible. However, newsworthy items from the Federal Government generally are not that a product is safe. It is more newsworthy to say that a product may be dangerous.

What also has become newsworthy in the regulatory area are disputes among the agencies. In our democratic process we have discussions among agencies and we do not always agree. A remarkably clever way to deal with your disagreement is to go to the press, and agencies start accusing each other of trying to do too much or too little regulation. Trying to resolve some of our differences publicly tends to slow the process down, creating some of the problems that Roger Blobaum was talking about. How do you resolve these issues?

I think part of the frustration that we see relates to the question from the Center for Science in the Public Interest—what is the level of regulation? Are we doing enough or not enough? I think the agencies themselves are trying to do the right thing. They are, in fact, all trying to cut back on regulation because they know they are overregulating biotechnology, either in the plant area, the pesticide area, or the food area, and they do not think it all is appropriate for federal regulation.

MR. URBAN: I do not presume to speak for farmers, particularly with co-ops here. I do, however, spend a lot of time in the field with farmers, talking with them, trying to be sure that I keep abreast of what my customers want. They feel in need of friends. The farmer today feels beleaguered. He does not understand the regulatory process; he is afraid of it because he sees the regulatory process as driven by consumer attitudes and environmentalists. He perceives that a potential outcome of that progress is to put him out of business.

So on the one hand, farmers are very concerned and feel somewhat powerless, faced with the normal pressures of business and consumers. On the other hand, they are also very concerned with the information that they are getting about their historical practices. They are not happy that they have been exposed to chemicals over the past 30 years. They are feeling that they may have been betrayed to some degree by those of us on the input side who maybe did not tell them the whole story (although probably we didn't know it ourselves). There is a real concern about what has happened to them, their own personal condition over a long period of time, and what is going to happen to them in the future.

These important links in the food chain feel themselves to be at the bottom of the barrel. They feel that other segments of the food chain, from the consumer all the way back to the input people, have more power in the food chain than they have. That is not a new feeling for farmers. Those of you who understand farmers recognize the populist impulse. But I would say it has been significantly enhanced by the kinds of experiences that they have been going through in the last 15 or 20 years.

MR. HETTINGA: We spend a lot of time with our farmers, both communicating with inputs as well as providing information to them. There is a great deal of concern out there that the consumer does not understand them. We try our very best to stay in constant touch with the farmer for those inputs, to get those concerns into the marketplace as well as to explain the difficulties concerning biotechnology and all other technologies that affect the farming community.

Dr. Miller mentioned that we have to improve the competence of the FDA and the regulators in Washington. I also think we need to strengthen their confidence. They get attacked by Congress and in multiple other areas, and they will quickly say no because of insecurity.

MS. CULLITON: I was interested in Tom Urban's remark that he spends a lot of time with the farmers and that he can almost now identify certain farmers with certain products or shelf space. Would you elaborate on that? Give us an idea of what you mean and what kind of technologies you are talking about.

MR. URBAN: You can identify poultry growers today with Tyson, or you can identify them with ConAgra. They are growing a highly modified product in terms of genetic uniformity, genetic feeding systems, sizing, and time of harvest of the crop. All of that ties a farmer directly to the food system.

The same thing is beginning to develop in the pork business. One of the problems of IBP, for instance—which is, I think, the largest slaughterer of hogs in the United States—is that their average lot size is 30 hogs. That is a problem if you are trying to run a huge, rapid slaughter operation. They are in search of a product that has uniform characteristics, making their operation more efficient, and delivering a uniform product to the final consumer.

So you are contracting with the farmer to do certain special things that you want to deliver to the consumer. The cattle business is moving the same way. We also are seeing it in the grains. Whether you are growing food corn for Frito-Lay, or high amylose corn, a specific amino acid profile in soybeans, you have a farmer who is responding directly to consumer needs through the distribution system, which is an enormously different process than farmers producing No. 2 yellow corn for their commodity trade. We now have large food companies beginning to recognize that they have to, in some way, relate to farming operations. They simply cannot go to the market and buy their input. They have to begin to affect the production habits of farmers.

The social implications of this are rather extraordinary. In agriculture, there is a Jeffersonian tradition that says that I am beholden to no one, that the moral fiber of this country rests with the man on the land, and that contracting or "kowtowing to the needs of the food system" is, in fact, contrary to that historical philosophical position.

That is shifting. The farmers are beginning to recognize that their access to technology and their access to capital is enhanced by participating in the food system.

There is a very real struggle going on in middle America between those two philosophical traditions.

The farmer, however, is beginning to recognize that, unlike the laborer who has only the laborer's work to sell, he has labor, technology, and owns the land and buildings in many cases. So the farmer has an income with a maximum/minimum return through contracting, but also unlike labor, has capital formation going on. That is not a bad world to be in. The 300,000 farmers who now produce 80% of all of the food and fiber in the United States are coming to understand that they are every day in every way more integrated into the food system.

MR. TAYLOR: Certainly in poultry breeding, the breeder has now entered into a dialogue with the processor to understand his needs in terms of white meat, body conformation, meat-to-bone ratios, and so on, in addition to the traditional productivity and food conversion characteristics that have always been a part of the breeding and selection process.

But I think it goes a bit further than that. Speaking as a neo-Luddite, I am glad the discussion is moving away from new technology and modified technologies and the regulation thereof, into what I think are equally important issues further down the food chain, the bottlenecks and the impediments.

In Europe we are seeing a situation where the dialogue really goes beyond the dialogue between the breeder and the farmer, or the breeder and the producer, or the producer and the processor. It goes right through to the guy who is actually selling the product to the consumer. This was exemplified to me very clearly about two years ago, when one of the most active participants at our Nicholas Turkey Breeding Conference was the man from Marks and Spencer, that is, the food retailer at the other end of the chain.

I think there is a lot to be done to get this chain to work well and to move the signal from the consumer back to the food processor, to the farmer, and to the geneticist. I think we ought to be paying more attention to the distribution end of the food chain. We have already paid a lot of attention to the other end, the supply side. A lot of the food problems that we see in the world, not in the United States, but in Central and East Europe, and in Africa and Southeast Asia, are not related to supply problems. They are not related to farm problems. They are related to distribution problems. I hope we will get some attention to that in the course of this conference.

PROFESSOR GOLDBERG: One of our other panelists who was supposed to represent the producer was unable to be with us. That was Dr. Kurien, who represents six million farmers in India, and who made a major contribution by applying technology to landless laborers, small-scale producers, and medium-size producers in that country. He took a system that was poorly run by the government and turned it over to the farmers, who created their own cooperatives at their local level and went through the whole food chain. They created a brand product for the dairy system that is the leading brand in India today, and did so by gaining the confidence of the farmer through the veterinarians who improved the technology. Dr. Kurien is also the first one to use BST in India to further that process along.

MR. BLOBAUM: I think that there is a certain alienation and lack of understanding between farmers and consumers and environmentalists. I had a recent experience that

I wanted to share with you on this point. A retired dean of agriculture is going around the country doing interviews on this whole question of dialogue or lack of dialogue, particularly between consumer and environmental representatives and farm and commodity organizations. I spent some time with him several weeks ago sharing some of my experiences.

I had a letter from him this week. He said, "I am nearing the end of my study and I have come to the conclusion that there is almost no attempt to have dialogue on the part of farm and commodity organizations." I would suggest that among my colleagues there is a desire for dialogue. I think that this is an unrealized opportunity that might greatly improve the situation that brings us together for this meeting.

MR. VAN ZWANENBERG: What is technology? It is a mixture of science and art. There has been a great deal of talk at this conference so far about science; not too much talk about art. In our experience, the customers have accepted that science has produced food in volume. What they want now is the craft put back into the food, and some degree of flavor acceptability. When we talk about poultry, our customers today accept that they can have poultry every day of the week if they want it. In fact, it is one of the biggest raw materials in our business. But our customers are asking for flavor. The fact that we can produce a chicken in 42 days is of little interest to them if it tastes like cotton wool.

What has science done for the consumer? In their eyes—in the United Kingdom at least—it has badly affected the environment. They have seen the creation of big scientific production units, and now they are being told that as a result of their activity, the world is polluted, producing huge quantities of waste that cannot be disposed of. We have to take food production from the beginning right through to the end. So far in this conference we are talking about production; we are not talking about the end products of production in terms of waste and the environment.

We tend to look at things with too much of a fragmented approach. We can look at the entire food chain and work together to produce things the consumer wants. There is an excess of food in Western Europe. We are putting it to waste. The consumer does not understand why. As a team effort—and I agree that there should be farmers here—we need to find a way of coordinating our activities so that we can affect the public perception, not in a fragmented way, but as a total food chain.

PROFESSOR GOLDBERG: Would you be willing to carry your ideas even further in a formalistic way? Would you be willing to take a leadership role? I might add that at our senior management agribusiness seminar that we have held for the last 32 years at Harvard, I have had the most difficulty getting membership from the retailing group. You are one of the exceptions; you are there most of the time. But from my experience it is only recently that retailers have decided that they are part of the food system.

MR. VAN ZWANENBERG: There is a more coordinated approach among the major retailers in the United Kingdom on a number of issues affecting the consumer. We are happy to participate. When I look at our American operations, if I look at yesterday's results, I am not so sure we are so happy that we bought them.

DR. VAN DER HEIJDEN: In the Netherlands we had one of Europe's biggest tomato industries. At groceries, they were appreciated because of their beautiful exterior. They had, in addition, extremely good storage and transportation qualities, but unfortunately no taste. That made the acceptability by the public almost zero. It was not biotech; it was just breeding and the wrong selection. Natural selection or biotech is basically the same process. The technology destroyed the capacity, at least in Holland, for the tomato industry. So the best you want to buy is now from Italy. Their products were selected for taste.

PROFESSOR HIGGINSON: It would appear to me that the role of the scientist relates to two entirely different areas. One area is the technology related to improved production, quality, safety, pesticide-resistant plants, and so on. The second involves an entirely different group of scientists, who are unconcerned with food production but who regard many of these technical improvements as potential hazards to society. It would appear to me that those two issues should be carefully separated when considering the role of the scientist.

Secondly, the implication has been made that if the public perceives the new technology to be beneficial, it will be easily accepted. I would respectfully submit that this will vary. If one is dealing with a technology to increase production in a country where food supply is borderline, the public will readily perceive benefits, as in the Green Revolution. On the other hand, the situation in this country is that the color of cherries, or the taste of a chicken, become the issues. The public seems to feel that, in the days of food additives, a greener green is almost automatically dangerous. If looking at the global picture, you must distinguish between those that are already haves and those that are on the way to become haves.

DR. JACKSON: One thing that I think we should not pass over is that one very large group of consumers who have benefited from new technology and changes in the global food system, certainly in America and in Western Europe, is women, because new technology has given us time. The final foods that come onto the market in cook-chill form or whatever have given us time because we do not have to spend time in producing it. All I do is switch on the oven and put it in. In terms of selling the benefits of new technology to the wider public, this is something that needs to be brought out, because it is going to be more and more important.

The second point I would make is this: within Western Europe, a lot of the opposition to new technologies has come in recent years from its most prosperous member country, West Germany, as it then was. It was, after all, in Germany that the Green movement had its beginning, the environmentalist movement. The Green movement, now spreading to France and Italy, has been in the vanguard of those questioning whether we should have "no risk" assessments, whether we really need new technologies; it has opposed food irradiation, the licensing of BST, hormone growth promoters, and so on.

It would be interesting to know whether this opposition now exists in East Germany because it has suddenly been given all of the benefits of Western European and American industrial technology, presumably coupled with a reasonably flourishing green movement. What do the East Germans think about all this? Do they now grasp the opportunities of new technology, particularly in food production where their agriculture was very poor? Do they grasp that with both hands, or are we seeing a reluctant attitude toward it on behalf of East German consumers?

PROFESSOR SCHELL: Right now in East Germany we are witnessing people concentrating on other problems, such as trying to keep their jobs and making sure they extract major benefits from their new association with the West, rather than having a really deep concern for the problems you mentioned. This is even true for academics, I am afraid to say. Let me illustrate this with an anecdote. When confronted with the hard reality that Institutes belonging to the former Academy of Sciences of the DDR would be closed or would have to look for new tasks, some of their scientists proposed that they handle work with genetically modified organisms, since this seemed to cause problems in the West and there were no inhibitions in the East! These people did not realize that since they now belonged to one and the same country, they were also subject to the same laws and regulations. They thought that they would be able to do it the way they were used to under the previous regime!

MR. TAYLOR: I want first to say a word to Marks and Spencer. Eighty or 90 percent of the flavor characteristics of poultry comes from the feed, not from the breeder. I speak only as the breeder. The flavor characteristics lie with our customers, but we are all going to work together to get a much nicer tasting chicken.

I think there is a group of green issues that occur in the old East Germany and in parts of Central Europe where there should be no conflict at all, because often environmental abuse equates with poor business practice. It applies in agriculture in the overabundant and inaccurate application of fertilizers and agro chemicals. There is enormous scope for reducing usage without sacrificing yield. It occurs in the waste of energy that goes on in many industrial plants, and it applies above all in spillage and process waste, that again are environmental hazards and are just plain inefficient.

I think in these areas there is no conflict. But when we get into BST and such, I am going to leave that debate to others.

MR. HOFSTAD: I think that the farmer's attitude has changed significantly. Recall the last of the 1970s, with the fence row to fence row mentality and the resulting overproduction. Then we had some critical events, for example, embargoes, and the recession in 1979, 1980, and 1981. I think the farmer of today recognizes he is a part of the food chain and it is not practical to go to the government to assure himself of a price for his production that will keep him on the farm.

The farm organizations in the next 10 years will have less and less influence on farm policy. Government will look to those people and organizations who are more actively involved in the production and marketing chain. Therefore, I think the commodity trade organizations and the cooperatives are going to have more influence if they recognize that they are part of the food chain and are responsible. We are no longer in the feed business; we are in the livestock production business. We are no longer in the chemical fertilizer business; we are in the crop production business. We were one of the first organizations to strongly support licensing people who apply chemicals, recognizing the environmental issues—food safety, human safety, and clean water.

If those are things we need to do to get more acceptance, let's get it done. Let's find out where the fears are and do something. We have 1500 local cooperatives in the Midwest, and they are ready to say that, if there is a legitimate concern, let us license those people so we can minimize these kinds of concerns by the consumers. Ten years ago Land O' Lakes organized a consumer advisory committee; not a farmer advisory committee, a consumer advisory committee. We listened to them intently and they influenced decisions.

I am optimistic about the future and our ability to listen and respond. Don't look to the Farm Bureau, Farmers Union, the Grange, or other farm organizations to be the leaders influencing farm and food policy. Look to cooperatives and the commodity organizations, corn producers, soybean association, pork producers, and so on—to the leaders.

PROFESSOR GOLDBERG: With your permission, I would like to try to summarize what my colleagues have said up here, and what you have said in the audience.

Professor Schell said that for better or worse, science is equally dependent on industry in its application and its interpretation of where it is going, and the industry-government-science partnership has to work more closely and more effectively if it is going to make sense to the scientists.

John Cohrssen echoed the same sentiments, but he said it is an industry-government kind of interface and that as one of the last countries of the world to have industry-government cooperation, we are long overdue for it.

Roger Salquist felt that we are making tremendous progress, but we have impediments and misunderstanding, and we have to break down those barriers.

Tom Urban indicated that the whole food chain has become an industrialized system, and that system itself will help us muddle through our problems.

Jonathan Taylor said we have to be most conscious of the consumer, and the partnering that is going on in the food chain in response to that consumer.

When I listened to the audience, I heard something else. I heard you say that things are better out there, that there is better understanding; that people are willing to accept the science, but that they want to see practical real results that are consumer-oriented. The time is long overdue for having a forum where this kind of group—with more producer participation, and more consumer, environmental, and nutrition participation—can begin to have the leadership and dialogue that the whole food system wants with the new scientific revolution. It has been one of the more optimistic sessions I have participated in for a long time.

PART II

REGULATING THE GLOBAL FOOD SYSTEM: HARMONIZATION AND HURDLES

Caroline Jackson

INTRODUCTION

The Uruguay Round of the General Agreement on Tariffs and Trade (GATT) is leading the world into a new era in the setting of international food standards. GATT aims to reduce protectionism by progressively removing trade barriers. Securing a sufficient supply of food for the world's population is a primary objective of many governments. Of all the individual agreements made under the GATT, however, those relating to agriculture are causing the most problems and have contributed to the failure of the current round of negotiations. Although the level and acceptability of agricultural subsidies have resulted in failure to conclude the GATT negotiations, these may in the long term prove to be minor hindrances compared with the difficulties that may arise over trade in food. Ensuring that food is safe, of acceptable quality and that it meets dietary needs are also high priorities for governments.

While the amounts and types of food produced and the nature of agricultural support vary from region to region, so does the range of legislation relating to agriculture. Both science and politics play major roles in the food regulatory process from the farm through all the stages of production to final consumption. Different countries adopt different methods to assess the safety of foods for their population. Account is taken, for example, of different dietary habits, climatic and environmental conditions (which can influence the use of pesticides and preservatives), or a society's view on acceptable levels of risk. Thus, whereas one country may allow the use of a particular additive or pesticide, another may not. Differences in food policy objectives and regulations create barriers to free trade, hinder economic growth and increase costs. The result has been trade wars, for example, the European Community's ban on the use of BST or the United States' temporary refusal to admit EC wine containing residues of a pesticide, procymidone, that was not registered in America.

To overcome obstacles to trade, consideration must be given to the ways and means of effectively regulating the global food system to reduce disparities while maintaining acceptable levels of consumer protection. This chapter indicates some of the hurdles that have to be overcome, examines the American and European regulatory systems, and considers ways of harmonizing existing systems.

DIFFERENT APPROACHES TO FOOD LEGISLATION—THE AMERICAN AND EUROPEAN POLICIES

The American System.

Responsibility for food policy in the United States is divided among a number of agencies. Food safety is under the control of the EPA (Environmental Protection Agency) and the FDA (Food and Drug Administration). The U.S. Department of Agriculture regulates meat and poultry. All pesticides, additives, and substances coming into contact with food have to be licensed by the EPA and meet current scientific standards. The EPA has responsibility for establishing the maximum legal limit on residues in food. It incorporates a "risk-benefit" strategy.

The FDA is responsible for the premarket clearance of new additives and new uses for existing additives. It maintains a list of safe food ingredients, regulates food additives and is responsible for enforcement. The U.S. regulatory system is essentially science based, and substances must meet stringent conditions before being licensed. The FDA considers safety, but may not take need into account. The work of the two regulatory bodies is subject to the Freedom of Information Act and so has to be made public and can be scrutinized by Congress and the courts.

European industry sees the American system as free from the conflicts and confusions that arise among the various bodies involved in the production of EC legislation: the European Commission, which is responsible for initiating legislation and then for administering its execution; the European Parliament, which debates legislation in draft and can amend and hold it up—giving a public voice to consumer and industry pressures; and the Council of Ministers, which finally adopts (behind closed doors) EC legislation, and which is prone to disputes and delays as national interests are—or are not—reconciled.

Consumer groups in Europe quote FDA decisions and see them as useful models. In the United States, the FDA seems to be criticized by industry for handicapping American competitiveness by being too slow and stringent and by consumers for failing to control food contamination and ensure high safety standards. Its independence has been questioned, and budgetary restrictions have been used to limit its effectiveness.

Food Policy in the EC

The EC is still in the process of developing a common food policy that will lay down safety requirements and facilitate the removal of internal barriers to trade after 1992. In the past the Community sought to establish compositional standards for a wide range of foods, but this approach failed because of widely differing perceptions and the desire for choice. Some compositional requirements, such as those for jam and chocolate, remain. The new approach, however, is to create framework directives and regulations that establish general safety and labelling requirements (horizontal legislation). More detailed requirements for particular food sectors and for special problems also are being drafted. Community policies are being developed to reduce the risk from animal diseases, to help control food poisoning, and to limit the spread of food-borne disease. Pesticide approval will be subject to a common system and maximum residue limits for permitted pesticides on all crops gradually are being developed. Chemicals used in and on food must be assessed for safety and conditions of use are being established for all permitted additives. A review of contaminants and processing aids will follow.

To reduce the duplication caused by multiple national safety assessments concerning chemicals used in and on food, novel foods, new processes, and microbiological contamination, the role of the Scientific Committee for Food—the group of independent experts that advises the European Commission—is being expanded. At the same time, Member States will divert resources away from national assessments into an expanded Community scheme. A draft directive to achieve this redistribution of resources has just been published and if adopted will come into effect in 1993.

A major weakness in the Community approach so far, however, has been the failure to adopt a general Food Hygiene Directive for which the European Parliament has pressed, although unofficial texts were circulating in the summer of 1991. Sectorial hygiene standards are likely to be delayed until the necessary framework legislation is published. But are separate Community rules necessary? Food-borne disease and recorded cases of food poisoning are increasing throughout the developed world. Could international regulations adequately protect the public and reduce the risk to public health?

Whatever legislation applies, it must be enforced. Common rules concerning enforcement and inspection have been adopted in the EC, but responsibility for enforcement rests with competent authorities in individual Member States. There is general concern that different standards of enforcement in Member States will render the Inspection Directive ineffective. Attempts are being made to improve coordination between enforcement agencies, but if this fails, the Commission has threatened to set up a Community Inspectorate.

Indeed, concern that controls on foodstuffs after 1992 will not be effective has led to calls from some Members of the European Parliament for a European Food Agency or Unit equivalent to the US FDA. This would be an independent central body to:

- coordinate the work of the Scientific Committee for Food
- operate a centralized, coordinated policy for acceptance and regulation of new products, additives, and processes throughout the Community

- be responsible for a Community Food Inspectorate
- operate as an open, accessible, decision-making process independent of industrial and agricultural lobbies, perhaps with a bias in favor of the consumer in cases where scientific doubt arises.

The proposed reform of the Scientific Committee for Food is a partial response to this demand. The aim is to allocate specific substances to an individual Member State for evaluation. The rapporteur, responsible for reporting on the substance in question, will come from another Member State in the EC and will submit a report for adoption by the SCF. It is not clear how the evaluation of novel food and nutritional and dietary questions will be dealt with under the new arrangements. The intention is that all Member States will conduct some assessments on behalf of the SCF in place of national evaluation of all substances. Some Member States will use existing scientific committees or research institutes to carry out the work. Others will use consultants as they do now. The effectiveness of these changes will depend on the commitment of additional resources from both the EC and Member States, and support from national scientific agencies.

Differences Between the Two Systems

i. Risk Assessments

Because there is no absolute guarantee of safety, because toxicology is an inexact and immature science, and because knowledge of the long-term health effects of consuming different foods is limited, risk assessments form the basis of most safety evaluations. The Americans have agreed on a list of substances that are generally recognized as safe (GRAS) on the basis of long established use. They also place considerable emphasis on carcinogenicity assessments because of the Delaney amendment, which seeks to limit the risk of cancer. This is interpreted by the regulatory authorities as an estimated greater than one extra case of cancer per million population. In the EC's Scientific Committee for Food and the WHO/FAO Joint Expert Committee on Food Additives (JECFA), no effect levels are identified following animal experiments, and depending on the nature of any adverse effects found at high levels of consumption in those experiments, safety factors are established to limit human risk. These determine the acceptable daily intake (ADI) that people can consume over a lifetime without exposing themselves to unacceptable risk. The FDA also uses the concept of ADIs, although it is not used as a basis for risk assessment as in every EC country.

Because differences in risk assessment and available data lead to different conclusions over safety, public confidence in the science-based approach has been weakened. Governments are looking for absolute results on which to base regulations. Some believe that factors other than safety should be taken into account; for example, The Netherlands government now combines a conventional risk assessment with a public appraisal to determine whether a substance may be used. Conventional toxicology is not appropriate for the evaluation of whole foods, as opposed to chemical additives and contaminants, and new methods of risk assessment will have to be developed to assess the safety of novel foods, new processes, and the dietary implications of changing eating habits. There is no consensus as yet on how this should be done.

ii.Need and Socioeconomic Benefit

In the United States, substances for use in food are assessed entirely by staff. There is an element of need assessment, but the way in which this is done, and the criteria on which it is based are unclear (at least to some of us in Europe). There is no formal and independent assessment of need, although additives must perform a specific technological function in food. To many industrialists in the EC, this is an advantage, but is opposed by environmentalists and consumer groups.

Some Community countries, such as the United Kingdom, aim to limit the addition of "unnecessary" chemicals to food by incorporating a criterion of need in the evaluation of additives and pesticides. In the United Kingdom an independent committee made up of representatives of industry, retailers, enforcement officers, and consumers assesses the need for each proposed substance. This has been built into EC additive assessments by means of a criterion of consumer benefit. In the United Kingdom, consumer benefit takes account of:

- the need to maintain the wholesomeness of food products
- the need for food to be presented in a palatable and attractive manner
- convenience in purchasing, packaging, storage, preparation and use
- the extension of dietary choice
- the need for nutritional supplementation
- any economic advantage

Annex II of the EC Food Additives Framework Directive* sets out general criteria for the use of additives. They can only be approved if

- there can be demonstrated a reasonable technological need and the purpose cannot be achieved by other means that are economical and technologically practical
- they present no hazard to health
- they do not mislead the consumer

The use of food additives may be considered only where there is evidence that the proposed use of the additive would have demonstrable benefits to the consumer; in other words, it is necessary to establish the case for what is commonly referred to as "need."

"Need" is not solely a matter of function, of manufacturer's assessment of technological need, or of new substances duplicating others already on the market.

*Council Directive of 21 December 1988 on the approximation of the laws of the Member States concerning food additives authorized for use in foodstuffs intended for human consumption 89/107/EEC. Official Journal L40/27.

The extent to which consumer benefit is formally assessed in practice in the EC is a matter of some concern to consumer groups.

European concern to add another criterion, the "fourth hurdle," in the assessment of drugs, veterinary medicines, and pesticides arises from the same background of environmental pressure and the desire by extra-industry players, many of them politicians, to question the existing basis of industry's regulatory arrangements for biotech products. The EC's system for approval at the moment includes the need to satisfy three criteria—safety, quality, and efficacy. The arrival of BST just at the time when the EC was attempting to limit milk production was a propitious moment for the launch by Socialist politicians of the idea of a fourth hurdle—"socioeconomic criteria"—since even if BST were ruled to be safe, its opponents argued that it was not needed and would disrupt European agriculture. The use of BST is banned in the EC until the end of 1991 while the Commission undertakes further studies—which will no doubt be concerned precisely with its socioeconomic impact.

The attitude of the European Commission to the fourth hurdle concept remains blurred. Commissioner Bangemann has stated that there will be no formal fourth hurdle, but at the same time the Commission reserves to itself the right to apply criteria additional to those of safety, quality, and efficacy when circumstances demand it. This flexibility may be something that European consumers and consumer groups welcome, but it can create a very confused—and litigious—regulatory picture.

This attitude is already well illustrated in the case of EC legislation banning the use of artificial hormones in beef production. This was a political decision following consumer pressure and social considerations; the scientific evidence never supported it, and as predicted, the ban is proving extremely difficult to enforce. Only the British government stood out against the policy, which is now in force throughout the EC, and which has been the source of a lengthy EC/US trade dispute.

Whatever these technical objections, however, the fact is that the fourth hurdle is in play and the EC hormone ban is in place. "Socioeconomic criteria" may be a yardstick invented by European politicians, but it is a highly exportable idea and will have to be taken into account in any attempt to produce a basic pattern of regulation in the global food system.

Another area of divergence between Europe and the United States concerns the regulation of genetically engineered—biotech—products. The Europeans have adopted an approach that treats these as a special category; the FDA has not. There are some advantages in the European approach: the centralized system of approval for high-tech drugs produces quicker results than in the United States. But new EC rules on the use and release of genetically modified products are stricter and may be more administratively burdensome than those in the United States. The special treatment of biotech products in the EC is to a considerable extent the consequence of pressures from the environmental movement—the Greens—of the 1980s, when industry itself wanted legislation in place to protect it from worse Green excesses.

iii.Enforcement

Legislation is of little use if it is not effectively enforced. The United States gains advantage over the EC at the present time because the office for compliance and enforcement is part of the FDA. Officials from this department are responsible for applying one set of standards throughout the entire United States food industry. But how effective is this in practice? Does the FDA have sufficient resources to achieve uniformity of enforcement in practices? To what extent do state authorities support the FDA in this? There are some examples of state governments taking action against deceptive labelling, for example, where the FDA failed to respond to public petitions.

Lack of uniformity of enforcement is a problem for the EC. Member States retain responsibility and jealously guard their autonomy in this area. It is too early to say whether the Community controls over inspection will be effective, but the failure of some Member States to implement existing legislation in a number of areas gives cause for concern. If the EC cannot reach agreement on the need for a European Food Agency, is agreement on safety and enforcement on a global scale possible? Effective enforcement is another hurdle that must be overcome for the global food system to be regulated effectively, consumers to be protected adequately, and competition among companies to be fair.

iv.The Position of Exports

European Community food safety laws apply only to food traded between Community countries and not to exports. This raises ethical and practical questions. If, for example, the EC strictly limits the use of certain pesticides in food production, would similar limits be applied to foods intended for export? If certain substances are considered to be a danger to health in one region of the world, should they be used elsewhere?

Should the EC be more responsible in its policy toward people in other countries, and would a change in policy be matched by an equivalent commitment from the USA or the other major trading nations? Should concern with safety take priority over supplying food to nations that lack sufficient supplies to feed their populations? Is any food better than none?

Both the United States and European Community tend to control the substances requested for use in their respective areas. Food producers may be allowed to use substances on exports that are not permitted internally, which is morally unacceptable if there may be a risk to health. Economically, however, a pesticide required in, for example, North Africa may not be needed in North America. If the United States or Europe does not allow the residues of that substance to remain on food, a barrier to trade will exist that may not be justified on safety grounds.

Importing countries may wish to make their own assessments of what is or is not good for their own populations. Respecting an individual country's rights to interpret scientific evidence as they see fit can be a hurdle to regulating the global food system.

Can these hurdles be overcome in the global system by reference to a respected international body whose decisions then can apply universally? Could the Codex Alimentarius Commission fulfill this need? What changes would be required in its structure to gain international confidence?

The Codex Alimentarius Commission was established in 1963 to implement the joint FAO/WHO Food Standard Program. Since its establishment, Codex has formulated over 200 commodity standards, including some on additives and contaminants. It has evaluated additives and pesticides for safety, set approximately 2000 Maximum Residue Limits for pesticides, and drawn up 35 Hygiene Codes. Codex also is concerned with veterinary drugs and recently has become involved in assessing novel foods. Approximately 130 countries are in membership, but they are not bound by its decisions.

Its main purpose is to ensure fair practices in international trade and to protect the health of consumers. Many now doubt that it has fulfilled this role satisfactorily, but could it form the basis for an effective global food system?

THE EFFECTIVENESS OF CODEX

The developed nations are generally reluctant to adopt Codex's recommended standards, which are believed to be lower than national requirements. Safety levels are based on advice from joint FAO/WHO expert committees. However, the members of these committees usually represent only a very limited number of the member countries and often do not cover the whole spectrum of expert opinions available. Since scientific evidence is limited in many areas, regulators in a number of countries favor a cautious approach when approving processes or substances where doubt has been cast. If Codex is to become more credible, it needs to adopt this more cautious approach and take account of public interest and concern about food safety.

Many countries also feel that a full debate on "acceptable risk" and "need" is essential before any substance or process is approved for use and should be incorporated into Codex assessments. Consumer groups also criticize the lack of representativeness of Codex committees, the failure to include consumer participation, and the lack of transparency about the procedures by which safety decisions are taken.

Although the developed nations mostly view the Codex standards as lower than they would themselves accept, the developing nations have difficulty in achieving them. This major obstacle is a barrier to exports from developing to developed nations and another hurdle in fairly regulating the global food system. However, it is vital, on humanitarian and economic grounds, that developing countries be helped to raise their standards of quality and safety.

LOOKING TO THE FUTURE

Food safety scares and the increase in food poisoning are undermining confidence in food production and control systems throughout the world. In the past the GATT did not refer explicitly to food. The draft agreement on Sanitary and Phytosanitary (SPS) measures, if adopted, however, will only allow measures that are:

- necessary to protect human, animal, or plant life or health and,
- consistent with available scientific evidence.

The proposal will split responsibility for food, animal, and plant health among Codex, the International Office for Epizootics and the International Plant Protection Convention. This may raise problems of overlap and gaps. Should one organization be given responsibility for regulation of the global system? How can policies established in separate agencies be made consistent, transparent, and equitable?

The GATT proposals give Codex enormous potential power. However, while it is probably desirable that a respected international body set worldwide standards, Codex falls well short of fulfilling this role at present. It is feared in many quarters that proposals for "harmonization" would mean that Codex's lower standards would supersede national legislation and in effect lower health and safety regulations. The GATT proposals undermine the criteria of need that are considered important by several European countries as a sensible precaution where safety data are incomplete. Issues of biotechnology raise ethical issues of concern to many cultures that cannot be resolved by reference to scientific evidence. Pressure for economic growth and profit from the food industry may conflict with desired dietary change and consumer benefit.

How can these conflicting questions be addressed? Is it necessary to do so? Will the new proposals benefit the developed nations at the expense of the developing? Can Codex be reformed to take account of these and other public concerns? If Codex is to become the international standards forum, what changes are necessary?

Consumer groups are campaigning for changes in Codex to increase consumer participation, improve the transparency of decision making, extend the input from experts throughout the world, and reduce the time for decision making so urgent safety issues can be urgently addressed when the need arises. The emphasis on international food standards poses a dilemma for consumer groups. They tend to oppose the covert use of national standards as protectionist barriers, but at the same time they do not want international standards to result in a reduction in consumer and environmental protection or an increase in risk from hazardous goods. The current procedures for determining safety do not meet consumer concerns so it is arguable that although greater harmonization of standards is desirable, total harmonization is unattainable. The GATT rules may have to recognize that contracting parties will find it necessary to introduce different standards in particular circumstances and that these may be based on wider criteria than scientific assessment of safety. Transparency and respect for GATT principles of nondiscrimination may be more important than uniformity.

There is a move internationally toward improved labelling in place of food standards as a basis for allowing food choice (as is happening in the EC). Increasingly, consumer definitions of quality are related to the degree of processing and the treatments undergone by foods, and it may be that comprehensive labelling, agreed definitions of certain foods, and methods of processing could provide a satisfactory alternative to full compositional standards for food.

Whatever changes are introduced, however, the GATT lacks a mechanism for determining whether individual countries' complaints about trade barriers are justified or not, and the case for compensation if a complaint is upheld. Such discussions will have to be considered in a broader context than scientific assessments of safety in use. The International Organization of Consumer Unions has criticized the proposals for their naivete about the validity of scientific assessments, undermining a country's right to set its own standards in response to consumer pressure, ignoring consumer interests and needs, and not protecting those in developing countries.

CONCLUSION

While the EC is committed to dismantling internal barriers and establishing food regulatory systems that can achieve this, there is evident concern across the Atlantic that the removal of internal barriers will be replaced by external constraints, and Fortress Europe will become a reality. The 1991 Community Programme expresses a commitment to "continue the process of opening up to the outside world," to relaunch discussions over the GATT, and to aim for trans-Atlantic solidarity among the Community, the United States, and Canada. If the conflicts between the world's biggest trading blocks are not to increase after 1992, there will need to be increased understanding of the objectives and methods of the different regulatory systems. Fundamental changes will be necessary to strengthen the global regulatory system and to ensure it has the confidence of regulators, industry, and the public.

The Codex Alimentarius Commission has the potential to become an effective world regulatory authority, supported by relevant national and regional authorities. This will not replace regional and national assessments, but could reduce duplication, and extend coordination and cooperation. To achieve this objective the Codex Alimentarius Commission must be strengthened. More liberal trade must be underpinned by high international safety standards that are enforced effectively. The procedures for evaluating safety should be reformed to reduce duplication, improve the scientific basis for evaluation, and make the systems more open and accountable to the public which they purport to protect.

CHAPTER 6

Science and International Food Regulation

Kees A. van der Heijden

I am convinced that we are moving toward global regulatory harmonization, albeit slowly. In analyzing the main problems in harmonization (and in Europe we have considerable experience in dealing with the various countries), it is clear that there are a number of players in the game. These players should never be mixed. The politicians and scientists and other players have their own responsibilities and we should keep them clearly separated.

First, in Europe there is the political mandate to harmonize given by the Commission. Second, there is the scientific part of regulation and its resources. Good regulation starts with good science. Scientists should be completely independent of politics or of any public or private pressure. The third partner in the game is the civil service, or whoever is responsible and has the mandate for legislation, regulation, and the political consensus. The fourth player, separate from the previous three, is the agency or agencies responsible for control, inspection, and law enforcement.

In addition to those four there are a couple of secondary players who have input at all levels. One is industry, with innovations and push for new products. There are lobbyists for industry at all levels, including the political and scientific. Industry itself has also done a lot of harm to legislation by advertisements of negative claims: healthy food contains no additives, no artificial colors, etc. On the other hand, there are also consumers and the green parties with their own demands, sometimes related to what we call chemophobia, the unfounded fear of chemicals in food. Also, there might be interference by politicians at all levels, sometimes for their own opportunistic interest. Finally, the media want to make news, and good news is never, never favorable.

There is certainly a need for further harmonization. In 1993 the EC is to join with the EFTA countries. That means 19 countries involved with 380 million consumers, involving 40% of the world trade. Not yet involved are the Central and Eastern European countries, but they are coming soon. In fact, many of these countries are now basing their food legislation on the EC legislation; their own food laws are already being adopted. Therefore, beneath the surface, all types of contacts and connections have already been made.

The next step is world trade. Trade barriers are often raised at the national level. If the EC had the comfortable feeling of being almost ready and having finished harmonization in 12 European countries, suddenly you find 19 countries and there is a lot of work to do again. And world trade means that we have to deal with Europe, America, South America, the Asian countries, and Australia. So I expect and predict that there will be many more meetings and committees dealing with global harmonization in the coming years.

The vision and final goal should be that food be available and safe at all places on our globe, whether it is Washington, Brussels, Rome, or Hong Kong. It is nonsense to assume that a food additive or a certain level of a food additive might be more harmful in one country than in another.

Returning to the four parties of the game in Europe: there is weakness in giving the mandate to harmonize to the civil servants at the national and international level. They do not yet have a clear message or have the power to create harmonization, showing weakness on the part of the politicians. Therefore, the politicians and also the Members of Parliament in Europe should be provided with more authority.

The second step in harmonization involves science. The EC Scientific Committee for Food, consisting of international scientists, is different from the FDA or the EPA. All the members of the committee come from the member states and are, for the greater part, already responsible in their own countries for the job. That means we have an extremely skillful and experienced group of committee members. The mandate of the committee is no more and no less than to answer the question: Is it safe? If yes, to what extent? No ethics or politics are involved.

For the Commission, it is mandatory to ask the Committee about matters related to food safety and nutrition. The output of the Committee includes standards, conditions of use, recommendations in order to support the Commission's work, opinions, and risk-benefit information and communication to the public.

The Committee is trying to develop new approaches. Sometimes regulation is better or more appropriate when you are using the principle of Good Manufacturing Practice or Good Laboratory Practice, or doing it the same way as JECFA or the FDA. For example, the Committee should not give a figure for an allowable daily intake (ADI), but leave it free, depending on condition of use. In my opinion it is nonsense when you have a numerical ADI for a certain additive and you know that in the worst case scenario only 20% of the ADI is ingested by an extreme consumer.

The Scientific Committee cooperates with JECFA, a U.N. organization, and very often also with the FDA and EPA. I would like to emphasize that often science, toxicology, and safety issues are misused at the various other levels of the regulatory process by politicians, by regulators, and by consumers. But that is never the greatest problem. In my experience, there are no scientific issues regarding nutrition and food safety that could not be solved in a reasonable time in a meeting with international experts. The experience, whether it is JECFA or other national or international scientific groups, is that even very complicated problems can be solved when there is no influence of politics or any other pressure. Leave safety solely to the scientists;

it is their responsibility and mandate. The scientific evaluation is never the critical hurdle.

Consider the other three parts of the process: regulation, legislation, and politics. Leaders in these areas deal with regulations, the composition of foods, the various national laws, and also the interface with the producers. For example, they will have to decide the labeling issues. My point here is that labeling is not scientific at all and should be left to the regulators. It is a matter of need and justification. Never ask a scientist whether there is a need for anything: the scientist's skill is to answer questions of safety, no more and no less.

I consider regulation as the most critical component. It is most time consuming; years may pass before a certain food item is regulated satisfactorily. The main causes of this are lack of political control and lack of resources in the European Community, where a relatively small number of people are dealing with the whole task. In addition, there is a tendency toward bureaucracy, which again is the result of a lack of control by responsible politicians. Parliamentary control, including parliamentary power, is too weak. I consider this area as the main problem for legislation.

The fourth component, control, inspection, and enforcement—including the need to have certified laboratories internationally recognized by the United States and other countries—should and easily enough could be systematically and coherently organized. But there is a strong need for political pressure and the mandates to organize the system.

In conclusion, science is not the critical issue in regulation of either novel foods or biotechnology. We have proved in international scientific committees that we are perfectly able to handle the issues and to agree with each other on safety items.

In my opinion, throughout the world the process is not open enough. Consumer groups should be involved at the various levels of regulation, and especially at the scientific level. Very often, however, the industry reports are proprietary. As a result, it is often found that a compound has been evaluated by a scientific group, and then at a later stage, in the wrong phase at, say, the parliamentary level, consumers start asking to debate its safety again. That is not the appropriate place to do it; it hampers progress. Safety should be dealt with at the scientific level. So I would strongly suggest that consumers be involved early in the process and at the various stages, regardless of the differences that exist between Europe and the United States. In Europe we are more consensus-directed, and there is less direct confrontation among the various interested parties. The producers and media should also be involved in the early stages.

To achieve global harmonization it is better to move the regulatory process up to a high-level, neutral institution, certainly at least the scientific parts of it. Safety issues, the pure science for global harmonization, should be the responsibility of an institution from the United Nations like JECFA or WHO-FAO. That would be my advice for a further step toward global harmonization, because such an institution is basically nonpolitical, neutral, and able to make optimal use of the available resources.

CHAPTER 7

Regulating Biotechnological Products

Dr. R. Tsugawa

In order to attain a successful harmonization of food regulation of individual countries, Dr. Caroline Jackson has proposed that the Codex Alimentarius be strengthened and that high international safety standards, enforced effectively, be established. I agree with Dr. Jackson's proposal, and I think that the critical objective in this matter is to reach agreement on high international safety standards. Furthermore, these standards must be reached on the basis of accumulated scientific evidence. Mixing local special considerations with scientific standards must be avoided. Such local conditions must be solved politically.

Food regulation as related to biotechnology is an important issue. It is now widely appreciated that biotechnology has great potential for contributing to human health care, for improvement of agricultural products, and for solutions to environmental problems. We must try to develop this technology so that we receive the benefits derived from it. On the other hand, the safe utilization of this new technology must be assured. Debates about the safety of biotechnology have been carried out since 1973 when the technique of genetic engineering was established by Drs. Cohen and Boyer. The first guidelines were established by NIH in the United States in 1976, and those guidelines became a model for the experimental manipulation of genetic engineering throughout the world.

Discussions for the establishment of industrial guidelines for biotechnology were started by an expert committee of the OECD beginning in 1983. As a result of these discussions, the first recommendations for the industrial application of biotechnology were issued by the OECD in 1986. Discussions leading to more detailed guidelines for industrial applications of biotechnology are continuing and should be completed by 1993. These guidelines promise to be the most reliable for individual countries because they are scientifically based.

In Japan, governmental guidelines for industrial applications of recombinant DNA in agricultural products, foods, and so on, were established or are being established with reference to the OECD's guidelines. In the recommendations issued by the OECD in 1986, it was said about the state of scientific knowledge: "There is no scientific basis for specific legislation to regulate the use of recombinant DNA organisms." This implied that additional scientific evidence was needed to establish the specific legislation to regulate the use of recombinant DNA, and at the same time, that there should be no such legislation unless the scientific evidence proved the recombinant DNA techniques to be risky. Therefore, the Japanese government did not establish legislation, but established guidelines that did not have legislative enforcement. Further discussion by OECD has been moving toward some relaxation of its guidelines, based on the latest scientific information.

Discussion regarding the safety of foodstuffs produced through genetic engineering was started by OECD's expert members in June 1991. It is said that the discussion is moving toward the conclusion that the evaluation of safety be performed not on the process of genetic engineering itself, but on the products derived from genetic engineering. The EC Council's directives issued in April 1990, which are being drafted by DG XI and issued to the member countries, are said to be directed to the regulation of the process of biotechnology itself, in opposition to the trend of OECD's discussion. The EC's directive seems to be based not only on the scientific principle but also on considerations of social criteria, such as public acceptance. Furthermore, the so-called fourth criterion that has been discussed in the EC Commission clearly would be a matter of politics. Such considerations should be separated from the basic biotechnological regulation.

On the other hand, the U.S. government has maintained a coherent policy based on scientific principle and evidence, as is seen in its product-based regulation. Japanese industries support product-based and oppose process-based regulation, and they look forward to the activities of the EC's Biotechnology Coordination Committee, which was set up in March 1991. It is hoped that the Committee will revise the biotechnology-related directives of the EC in favor of technological development and the competitiveness of the EC.

In conclusion, I hope that the regulation of biotechnology, especially food-related regulation, will follow the recommendations that will be published in 1993 from OECD. This will be important not only for member countries of the OECD, but also for other countries belonging to the United Nations. I wish also to bring to your attention some of the activities of the Japan-U.S. Biotechnology Forum set up by the Japan-U.S. Business Conference. It provides an example of how to approach harmonization. The Japan-U.S. Biotechnology Forum was established by an agreement of the 25th Japan-U.S. Business Conference in July 1988. Its purpose was to identify and to reach agreement on fundamental scientific principles and procedures that form the basis of regulation and guidelines for biotechnology. More than 30 scientists in the field of biotechnology participated in this joint forum, and the effort was coordinated with related associations representing both countries. As the result of the two years of work, a report of carefully considered recommendations was delivered to both governments.

This activity has been a continuously developing one, and includes participation of the members of European countries and Canada as the International Biotechnology Forum. The activities of the Forum were carried out from the standpoint that the regulation of biotechnology must be established on the basis of scientific principle and evidence, as was the case with OECD. It is not based on economic or social factors, except for limited ethical problems such as the case of the treatment of the human embryo.

DISCUSSION OF PART II:

DR. JACKSON: We have tried to outline the difficulties that exist, both in terms of the problems that Dr. Tsugawa focused on—biotechnological products—and the procedural problems that Dr. van der Heijden focused on. We have highlighted a rather complex structure of regulation, certainly at the European Community level and we know at the FDA level. Is it possible either to improve what we have individually, and/ or build up international systems of regulation so that conflicts do not arise?

The central question in this discussion, as well as in Roger Porter's is: how do we reconcile or cope with both scientific evidence and political considerations in regulatory systems? I was interested to hear what Secretary of State Olechowski said about the comparative ease with which Poland would be able to cope with the whole body of EC legislation that has been passed and that any new Member State of the EC would have to take on board. If a country seriously believes that it can enter the EC and immediately have very little difficulty in complying with EC legislation—for example, in the food law area—then it really does need to consider its position carefully.

The national political interest is also a problem in terms of food regulation. Roger Porter discussed the conflicts that exist in the United States between what individual states think and what federal authorities think in terms of food law. In Europe there are conflicts taking place between national authorities in food regulation, who really still want to regulate the whole national market and believe that their methods are best, and the Community authorities. There is now a call to set up an EC food inspectorate. They see it with the role of monitoring and implementation, rather than a body centering on licensing and approval.

Can regional problems be resolved by reference to international bodies, to JECFA or Codex Alimentarius? It is interesting that recently Codex, which develops international standards for food additives and chemicals used in food processing, actually rejected the report of its committee on residues of veterinary drugs in food to the effect that growth hormones do not pose a threat to human health. Even if you look at the international level, the same conflicts are going on there.

DR. MILLER: First, I want to thank you, Dr. Jackson, for a superb paper. I have never seen anything anywhere that laid out the problems associated with the development of an international food safety system the way yours did. Agreed, it was a poem of questions; but nevertheless, I think they were the important questions that need to be resolved, many of which will not be resolved for a long time.

I would like to make a couple of remarks on three subjects. First, a brief statement about the structure of the food regulatory system. Second, a comment on enforcement and implementation, which, I think, is the key to any regulatory system. Third, a response to a comment that was made about transparency and participation.

Discussions of the food regulatory system become confused because we constantly confuse the issues of safety and regulation. Safety is a scientific issue, as Dr. van der Heijden said earlier. It is an issue that can be resolved by examination of the data and the application of judgment. If scientists cannot reach a consensus on a safety issue, then the product, by definition, cannot be considered safe. The problem occurs when that issue of safety becomes tied up with all of the other components of the regulatory process. It is perfectly appropriate in my view for political, economic, social, and cultural issues to be considered in the regulatory process. But they ought to be considered in their own right, and explicitly, rather than implicitly, hidden behind the screen of science.

The problem in my view with the hormone decision by the European Community was not necessarily that the community decided that it did not want hormone-treated beef. I disagree with it and I think it was a wrong decision, but I can understand their right to do it. What I resented was the fact that they did it behind the guise of arguing that hormone use was unsafe, and that is pure nonsense. The mistake is that we allow science to be used as a screen behind which all of those other things included in your fourth hurdle are considered. They have always been considered; but they have always been considered in private, not directly as an explicit part of the decision-making process. It is very important that any decision-making process involved with regulation separate the science from other considerations.

The other characteristic process of regulation-making is its adversarial nature. Scientific discussion and the reaching of scientific conclusions should not be adversarial. The issue of required participation by one group or another, by industry or public interest groups, does not belong in the scientific decision process. It belongs in the regulatory decision process, and, as I pointed out, there is a distinct difference.

Second, let me say a word about implementation and enforcement. I honestly do not understand how the European Community will assure the safety of foods passing through the borders. To do so will require assurances that the food production systems in each of the countries are meeting the same safety standards. I am not talking about standards of identity or composition or whatever; I am talking about hygiene standards and food safety standards. There is a tremendous difference among the 12 countries of the Community and their ability to provide inspection services. Without any kind of supranational monitoring, I think we will have to wait for the food intoxication incidents to occur. Are we compelled to wait for the bodies in the street before the conclusion is reached that a centralized inspectorate is required?

I realize it is naive to think that countries will readily give up their sovereignty on this issue. Nonetheless, at the very least, some kind of international monitoring is needed for any free trade agreement, whether it is here in North America, or in Europe, to assure that the same safety standards and health standards are being applied across the board. The Codex Alimentarius of the FAO/WHO can play a big role in this process. The Codex's problem up until now is that it really is a group of countries attempting to maintain their individual sovereignty and at the same time to gain or maintain some advantage in international trade. It is also underfunded for its mission. But it offers an opportunity to provide the international forum required for these trade issues.

I think the Joint Expert Committee on Food Additives (JECFA) is quite different.

Dr. Jackson made a statement in her paper concerning JECFA with which I disagree. I think JECFA can and does do the best job it possibly can in the scientific area.

Finally, I wish to comment about transparency and participation. The U.S. system is a transparent system; the nature of the Freedom of Information Act, the Administrative Procedures Act, all of the incredible number of administrative steps that regulators must go through in the development of regulations, will drive you up the wall. I once said to a group of government regulators that the only way in which I was ever able to get anything done at FDA was by sliding by some of these steps. It is impossible if you follow every single step to get anything done; it just takes forever.

Nevertheless, it is a transparent system and it does provide a number of opportunities for participation at different levels in the regulatory process. The difference is that participation in the process does not mean decision-making. It does not mean that there is a step called—in the terms of the Community—the fourth hurdle, socioeconomic issues. Nor does it mean that the regulation goes through some committee identified as being responsible for socioeconomic determinants and they, in turn, make a decision.

What it means, in the United States, is that everybody has an opportunity somewhere along the line to express a view on the regulation. In fact, in some cases the process can extend all the way to the Supreme Court of the United States. That is, I think, important because many of the European systems operate by fiat. As a regulator, I must say that I look with great longing on that process. As a citizen, I much prefer the system that we have today.

We ought to stop bashing the politicians in this area. I realize from me that is heresy. But I believe that when it comes to deciding the issues of politics, economics, and sociocultural inputs, they are the only ones who are ultimately accountable. With all due respect to my colleagues in the trade associations, the Center for Science in the Public Interest, Public Voice, and so on, they are not accountable except to their membership.

A politician is periodically accountable to the electorate. The politicians are the ones who, in the end, have to represent that electorate no matter how irrational that might be. If they do not make the decisions on politics and economics and socioeconomic issues, who will? Bureaucrats? Committees? I think that is an important consideration in the development of any system for assuring responsible regulation.

DR. VAN DER HEIJDEN: In the case of hormones, I agree completely. There was not even a scientific report of it; that was not requested. So that was purely opportunistic misuse of toxicology.

I am considering the controls, the standards, and the structure of, let's say, a future FDA-like institution in Europe. In my opinion, such an institution should consist of a very strong bureau of European civil servants dealing with legislation. In addition, there should be a centralized certified set or network of control laboratories. The most ideal would be if the present national laboratories of food inspectorates would join and agree to cooperate.

Regarding the scientific evaluation, I do not see a huge institution and large buildings in Brussels having all the scientific resources on safety and nutrition and the libraries together. So I think at present we still are dealing with an almost ideal situation, having the scientific and raw power resources from the various member states available. Instead of doing the work on a national level, the work is directed to Brussels. This means no duplication, optimal use of resources, and, far more important, it keeps science out of the civil servants' hands. A scientific committee stays independent from the civil servants, from the Commission itself. That provides the highest credibility to the public and the media.

DR. JACKSON: If I were an American whistle blower and looking for a new field to operate in, I would pack my bags fast indeed and export myself to the European Community because I suspect that there are all sorts of things that are going to go horribly wrong before long.

It is true to say that the Community system of food regulation as it will exist in the internal market after the first of January 1993 is really the triumph of hope over experience: the hope is that the methods by which food is produced and put onto the market in all the member states will be broadly similar and probably tending toward the highest standards. That has to be only a hope, because after the first of January 1993 there will be no rights to the infinite inspection services of the individual member states to stop imported foods at frontiers, to inspect them, and, if necessary, to turn them back.

It will be a system of random tests and checks by national inspectorates, and the Commission recently produced a directive on food inspection services that was so vague as to be almost not worth the forest that it cost to produce it. It included one memorable sentence to the effect that, within the European Community, food inspections should consist of "an inspection." There was no attempt in the original draft to include any reference to what the training of food inspectors would add up to. Only subsequently did it appear that training standards varied, from the requirement that food inspectors should in some countries, notably Britain and Germany, have a degree in food science or a related discipline, to two weeks training in Portugal. That was clearly something that the Community needs to address.

It is really under that umbrella that one moves toward the idea of a European Community food inspectorate. I think that this is going to come. The Community is moving toward a system of environmental inspectors who will be there to look after the inspection services of the member states, to breathe down their necks and make sure that they are doing the job themselves. That service will, I think, be created by the individual member states, acting together, within a fairly short time in the environmental sphere. I would think that by the mid-1990s, the pressure for a European Community force of food inspectors will have become irresistible—a force of inspectors answerable to an office back in Brussels, there to go around the member states to make sure that each is implementing Community law in the same way.

DR. ARNTZEN: It strikes me that we should deal with the irradiated food issue in regard to the public perception of food safety and the difference between politics and science. If we look at technologies that exist today that could be a major benefit to developing countries where food storage and transportation are a major problem, irradiation could be an important means of extending the food supply. It is an excellent means of preventing post-harvest losses due to microbial action. In many cases the developing countries look to Europe or the United States for standards in food safety. In the case of irradiated foods, if we have regulations against its use, it may be inferred that there is a scientific basis for not allowing irradiation. In fact, it is only a political compromise designed to respond to scientific illiteracy concerning the value of irradiation.

Is there some way to reveal to the public that there is this difference between science and political compromise so that in the case of existing technologies we might be able to break them loose, and utilize them where they are urgently needed? We should not let beneficial technologies stand on the shelf when we do not need to worry about them.

DR. VAN DER HEIJDEN: Food irradiation is allowed in Holland in certain cases. A new law has passed. It is not in the European Community as a whole for the time being.

The objections from consumer organizations and from green parties are not because they are afraid it is a health hazard. In the case of chicken irradiation, they think it is just a method to mask bad microbiological quality. They are convinced, providing there is good slaughterhouse practice, that there is no need for irradiation. Their suspicion is that all types of contaminated food will be irradiated and artificially refreshed. In addition, of course, there are politically-oriented groups who just do not want to deal with "radioactivity."

DR. JACKSON: I am in the position of having in my constituency the only plant within the United Kingdom operated by a company that has now applied for a license under the new British legislation on food irradiation to put irradiated food onto the open market. I have been around to see them.

It is quite instructive that they are trying to take outside groups around the plant as well, to show them what is going on. One of the largest lobby groups—in a way, I suppose it is equivalent to some of the American groups represented here—in Britain to take a great interest in food irradiation has the most innocuous name you could imagine. It is called the United Kingdom Federation of Women's Institutes.

Now they believe, as it were, to a woman, that food irradiation is undesirable for all sorts of reasons, encompassing all of the reasons that Dr. van der Heijden has given, including the superstitious one, that irradiation is somehow or other going to hurt them because it is connected with nuclear radiation. They also believe that it will be misused to mask contamination in food. Finally, they believe that, in fact, irradiated food is already entering the British market from Holland, but that we cannot tell because, of course, there is no way of detecting whether irradiation as a treatment has been used on food or not.

Now your question was, how do we get over the advantages of irradiation to the general public? Well, the only way that I and other politicians have explored is to point out that food irradiation is already being used perfectly safely within the United Kingdom and, indeed, in other countries as a treatment for food used in hospitals for patients who have to be on a germ-free diet.

That stops the United Kingdom Federation of Women's Institutes in their tracks. But it is not an argument that you hear very much. I suspect that is because of the fact that people do not like to admit the great increase in the incidence of food contamination, and possibly food poisoning, that has occurred, certainly in the United Kingdom, but also in Western Europe as a whole, over the last few years; and, therefore, to admit that food irradiation might be a useful tool in dealing with this problem. The problem may have arisen for a variety of reasons not least possibly through the poor cooking methods that people use.

Nevertheless, food irradiation is a difficult issue for politicians, and I do not think that we are being helped by the irradiation industry. Those of us who are prepared to promote food irradiation as a useful tool find that those who do the irradiating are silent when we call upon them to express their views as to the usefulness of the technique, which they actually put on the market.

DR. TSUGAWA: In Japan, except for the control of potato germination, irradiation for food products or food items is prohibited. Gamma ray, by the way, is used for potatoes in Japan. But I do not know the details about the scientific ground for that exceptional use, or the prohibition of the use of irradiation for other purposes. There might be some cases where you can use irradiation, including the gamma ray, for some positive applications.

My position is that we, first of all, have to identify clearly the scientific grounds for prohibition or the permission to regulate any use of irradiation. What I have just described is the opinion from, more or less, the industrialists' side. How to consider this issue as a political issue is a completely different subject.

PROFESSOR HIGGINSON: There has been an impression that it will be easy to separate science from politics. I believe this is wishful thinking, because politics is really risk management, or making a decision in situations where scientists may not have any degree of certainty as to solutions.

In the case of a big and obvious problem, there is not usually any problem in making a managerial decision. Where, however, no effect has been detected or is likely to be detected, there is usually a wide range of opinions for a number of reasons as to what "safety" measures should be taken.

Who then are the scientists on whom you are going to call? As a former director of an institute evaluating carcinogens, I used to convene expert groups. Quite honestly, I could select acceptable scientists who would give me the answer I wanted in advance. That is true of every committee. There are one or two scientists who run the committee and a chairman, and they basically will control the final results.

That agency is said to be probably the most objective organization in this context, as it is 3000 miles from Washington. But when coffee was recently discussed, they split and could not agree or disagree whether coffee is a human carcinogen. Eventually they had to agree that they could not agree and produced no evaluation. There is a common substance that has been around for a long time and about which numerous studies have been done. It is unlikely that further studies are going to solve the problem. There is an example of your having to make your decision as a politician, since it cannot be based on scientific consensus.

DR. GAULL: I was a little puzzled by one of the statements that was made. JECFA was described as nonpolitical. The implication was that JECFA and the Scientific Committee on Food were strictly scientific bodies, and we have heard a great deal of lip service given here to the universality of scientific decision-making. Yet there are innumerable differences of ostensibly scientific opinion between JECFA and the Scientific Committee on Food with regard to particular problems that each has looked at.

DR. VAN DER HEIJDEN: Recently, the same chemical was evaluated by three different committees in the United States and Europe: by FDA, by JECFA, and by SCF. I shall not mention the name of the additive. After the evaluation, there were three different opinions. But that does not show that they are all wrong because the margins of the difference were relatively small. On the other hand, those groups had not had a chance to talk to each other in the same room, so they had to come separately to their conclusions.

About the blue ribbon committees: there is one basic difference with the SCF. There is nobody in Brussels who can make his own choice or selection for inviting members for the meetings. Each member state will appoint the person already responsible in his or her country for it, so the selection procedure is completely different. There is no way to manipulate by inviting certain scientists.

DR. GAULL: Dr. Jackson, you have had some interesting experiences with differences among countries, concerning ostensibly scientific evaluations of food additives. What do you think?

DR. JACKSON: There have been differences. There may be different national priorities and national ways of looking at things.

A very interesting example in the sweeteners field at the moment has been the insistence by the British Government (and I think also by the French Government) that cyclamates should not be allowed onto the market in the United Kingdom. This is at variance with the overall view of the European Commission, and also of the Scientific Committee for Food on which British representatives have a place. I always find this a difficult subject to find my way through because when the European Parliament discussed this, we were told by a representative from the Commission that the Scientific Committee for Food always agreed. At that point we realized that we had stumbled upon the ultimate committee in the world: a committee that always agreed. But I am not sure that that actually was the case on cyclamates, because of British objections.

Nevertheless, what you see there is the reluctance on the part of a number of countries in the embryonic federal system in Europe to alter what is a national policy if there is a majority decision that goes against it. I think this means that individual member states will now have to reconsider their position on such matters. If they do not want to veto something or, even more likely, if they have the possibility of a veto constitutionally removed from them, they will have to learn how to live with the decisions of the majority.

DR. VAN DER HEIJDEN: I think this is exactly where politics or other nonscientific components are entering into the area. In this case, the Scientific Committee will never vote, and sometimes there is disagreement. This was the case for the U.K. representative. He was convinced that we made a mistake. But for France it was completely different in the Scientific Committee. The French representative agreed completely that cyclamates were fully safe, but there was a political argument later at a completely different level why France would not have it. I do not remember precisely.

MR. VAN ZWANENBERG: I think Marks and Spencer might be in a unique position to comment on some of the issues raised. We trade in six of the European countries of the EC. We have stores in Canada; we have stores in the United States. We, in fact, produce foods in 10 out of the 12 European Community members. We as a company believe in food legislation and harmonization, and we have worked closely with the British Government and with other governments to help that. I must say that in a general sense we think it is making good progress. The problem we have is about the harmonization of implementation in the EC, and it could be best expressed as a north-south issue—the north probably implements more than the south. It has always been said about the British Government that it "argues most but implements best!" Many of the other partners do not argue at all, but they do not implement the regulations either.

As to whether regulations can be implemented across the EC, all I can tell you is that in our experience producing in 10 countries, specification buying as we do, and having very strict rules on hygiene, safety, and so on, at Marks and Spencer we are managing to get a harmonized approach across all 10 nations in which we operate.

As to the separation of science from politics, if we live in the real world, it is not possible. I mean, the political issues are heavily involved with science, and we cannot separate the two, for example, in the case of irradiation. What is the real need for irradiation? It may be a scientifically proved method of preservation, but there is not actually any political need for it; not in the European Community anyhow. We can produce and preserve food and transport and distribute food to all 12 nations within the community without any need for irradiation. So politically there are no points to be scored or votes to be gained for it.

Can I come to the question of additives and ingredients, and to whether, in fact, something is safe or not? We do not add MSG to our foods. I know it is a subject that is going to be raised in a television program coming up in America shortly. You may say it is safe. Well, a recent report on food intolerance in the United Kingdom shows that a small proportion of the public suffers from what is known as Chinese Restaurant Syndrome. We as a company are not prepared to have one of our customers suffer from Chinese Restaurant Syndrome because that one person is 100% affected. It is not a one in a million chance, it is not one in a hundred million chance; it is 100% affected in the case of the individual sufferer.

So while I hear about food safety, I do not think you can remove emotion, politics, or anything else from science, because the public does not. We live in the real world.

DR. MILLER: To argue that because things are, that is how they should be, is a bit much. I might ask you whether or not you also do not use peanuts or chocolate in your products because there are lots of people sensitive to peanuts and chocolate.

The reality is that the use of these products is selectable by the consumer. You have made that decision because it is a business decision. That is perfectly proper to make; I have no argument with it. But please do not hide behind a screen that somehow or other you are doing something for the health of people. It is a business decision that you decided to make, not a health discussion.

Politics are part of regulatory decision-making. It is true you can pick committees that agree with you a priori, but that does not mean necessarily that you should. The fact is that politics are so deeply involved in the decision-making process in large measure because we permit it to be involved in the scientific process, as well as where it appropriately belongs.

If we do not know the answer to a scientific question, if we have dispute over a scientific issue, we must say that. It is at that point that the political decision on regulation has to be made. If you do not know and you cannot get a consensus from the scientific community, a reasonable scientific consensus, then it seems to me that a regulator has to say no. You must wait until there is enough data to reach some kind of consensus about the safety of the product.

It is possible to put together committees that do not come with a great deal of baggage or at least mix the baggage up so you can come to some kind of a decent conclusion if that is what you want to do. Food irradiation is a case in point. Whether the community needs it or not, to ban it because of some fallacious idea that it is unsafe is wrong. Every reputable scientific organization that reviews the data has concluded it is a safe process. The fact of the matter is, you have the right to ban it because you do not need it. That is fine. But do not argue that you are going to ban it because it is unsafe. The end result is that it reduces the credibility of the food system, it reduces the credibility of those who make the decisions concerning safety of the food system, and you end up in a much worse situation.

I happen to think that while politics does now play a role in the decision-making process of the scientific arena, it does not have to be so. We must begin thinking of putting together systems that allow those decisions to be made without the superimposition of things that are not concerned with the scientific decision.

MS. HAAS: I disagree. The food safety debate and the reason that there is a debate is because you have, on one hand, the economic interest of the food industry, and on the other, the health interest of consumers. Oftentimes that clashes. Where that clashes is in the public policy arena. You said we should not deal with how things are, but the way they should be. Policy should consider more than just the science because there is no absolute, as has been stated before. You can have "your" science or "my" science or "somebody else's" science; by nature, there is going to be a difference.

What we are really talking about is the acceptability of risk. When you are talking about the acceptability of risk, you have to consider the other concerns, the social concerns, the fourth hurdle, as it is called in the European Community. But we are dealing with the same things in the United States. We are talking about the acceptability, not the extent, of risk. The extent of risk is all that science can give us.

DR. VAN DER HEIJDEN: In the field of food additives and ingredients it makes no sense to discuss acceptability of risk, because basically the use of these chemicals is considered safe; there is no risk for the consumer. So it is confusing to raise questions about what might happen to one of a million. There is no risk at that level. That is the principle of the process. The only exceptions are carcinogens.

MS. HAAS: Again, I think that there are enough scientists who would disagree with you. In fact, there have been additives shown to be carcinogens that have been taken off the marketplace. You have additives that are interchangeable. When we got rid of one color dye, industry used another color dye.

MR. SHAPIRO: Mr. Van Zwanenberg said—and Dr. Miller accepted—that irradiation might properly be banned simply on the grounds that there is no need for it. I do not understand that as a matter of principle. It would seem to me the purpose of this regulation is admittedly related to health and safety. If there is no need for it, presumably the market will take care of that issue.

DR. JACKSON: I think the point is you are dealing with the question of whether or not you license irradiation as a process that can be used if people want to use it.

MR. SHAPIRO: I understand the point, but the argument was not made on the basis of health, and the argument was not made on the basis of safety; the argument was made on the basis of need. It would seem to me, therefore, that the appropriate result is to permit it and let people determine for themselves whether they think they need it or not.

MR. VAN ZWANENBERG: I was trying to explain where politics come into the question of food safety. The fact that irradiation is safe may be proven. In the United Kingdom it is permitted on certain foods, as it is in Holland. The question is whether there is any need for it politically, because within the European Community we can supply sufficient food preserved by other means. That is where the politicians have come in. There is no great incentive for the politicians to get involved in trying to permit irradiation.

DR. JACKSON: Mr. Van Zwanenberg, I think you need to go back home and ask in Marks and Spencer what treatment you are using on herbs and spices, because by the end of this year you are not going to be able to use ethylene oxide, and I strongly suspect that you will, in fact, be using food irradiation.

MR. VAN ZWANENBERG: We will be using heat processing.

DR. ARNTZEN: Ms. Haas, you made a statement I did not understand. You said it could be "your" science, "my" science, or "somebody else's" science, and it will all be different. The way I understand science, we work on a set of fundamental principles that are the same in the Netherlands or the United States or Japan or anywhere else. I just did not understand what you meant by the fact that we can all have different science.

MS. HAAS: If you take the case of Alar, there were scientists interpreting the data in different ways, on different sides of the issue. Scientists stated that there was absolutely no risk to humans, and there were others that recognized the risk. There did not appear to be any "absolute" answer.

I think that we continue to have controversy, as long as we continue to have uncertainty. If there were not uncertainty, we would not be having this debate and there would not be a political issue because there could be a decree that something is safe. Clearly, science is not at that point yet, and I remember Sandy Miller speaking at a Public Voice conference on food safety saying that risk assessment is black magic—he probably would still say it because I see his head shaking.

So if you still must rely on "black magic" for the scientific answer, then that is what was meant by the statement that there is your science and my science and somebody else's.

DR. ARNTZEN: Then I think we are not talking about science there, because I do not think scientists are going to disagree about what I call science; that is, is the instrumentation capable of giving an accurate measurement of how much Alar might be a residual in an apple? Is the toxicology standard set so that we can determine a level that causes 50% death or 50% rate of cancer incidents or something like that? But what you were referring to is an interpretation of whether the mass of data that has been accumulated means something in terms of public health.

MS. HAAS: Right. It is a public policy question.

DR. ARNTZEN: It is a policy question, but you were referring to your science and my science. I would go back to the argument that science is a standard and we can agree on what the scientific data is. It is when we get beyond that and relate it to policy and risk assessment that we come into controversy. I would hope that at least we would have come to the consensus that we do have a universal scientific standard that we agree upon, and we should not attack science or we should not hide behind science as presenting something evil.

PROFESSOR NELKIN: There have been studies of the role of scientists in disputes that relate the findings of scientists to their place of employment. In other words, estimates of risk are often related to the interests of a company or to political or professional identification. Estimating risk in areas of great uncertainty leaves a lot of interpretive space. This allows possibilities for political or economic influence.

DR. MILLER: In the first place, there is a distinct difference between talking about science and talking about policy. I want to repeat what I said before. Policy or regulation is a synthesis of a number of things in which science is only one part. I want to repeat that again and again and again. It is appropriate for a community to say we do not want this or to say we do want that. I do not care what the reason is. It is not appropriate for that community to say it is unsafe, or to imply that science says it is unsafe if, in fact, there is a consensus that it is safe. If, in fact, the scientific community can lay out a set of issues for which there is great controversy, the development of policy or the regulation has to take that conflict into account as part of the decision process. Science has done its job to the best extent that it can.

I think the mistake is that we permit people to select the science they want in order to prove a point. That is what I am objecting to. This is happening now and continues to happen. I object to that strenuously because I think all it does is reduce the credibility of science; it reduces the credibility of the policy; and it ultimately reduces the credibility of the food supply.

DR. TSUGAWA: We were talking about politics and science, and in my view science actually means evidence. The facts are the facts. But when it comes to interpretation of the facts, then here comes room for the people's will, desire, and so forth. Capturing the facts as precisely as possible is necessary, and that is what science can offer.

Today's session subject is regulating the global food system and harmonization and hurdles. When it comes to this kind of subject, the most important point is what kind of criteria we should use as the most reliable yardstick to formulate some regulations. In my view, science should be that most reliable criteria. Therefore, we have to share the common understanding about science as a common yardstick especially for the harmonization efforts.

DR. JACKSON: We have had a very interesting discussion. It is rather a pity that we did not move on to the question of the effectiveness of the Codex Alimentarius and other international standard-setting bodies as possibly coordinating regional regulators. I would draw attention, from the point of view of the European Community, to the very poor record of implementation by the member states of the food law which the Community already possesses, so we are actually discussing seriously something that many people are ignoring.

On the question of science and politics, I would only say that it seems to me obvious that, at the end of the day, it is the politicians who have to interpret science. But one thing that has not been said this afternoon is that it is very important that politicians should have an education adequate for that task. One of the problems that we are increasingly going to come up against is the fact that science as an understandable discipline has drifted away from the legislators and that, on the whole, politicians do not know very much about science.

PART III

INTERNATIONAL AND TRANSNATIONAL FOOD ISSUES

CHAPTER 8

Agricultural Trade and the New Polish Economy

Andrzej Olechowski

The question I wish to consider is this: What does one do with the agriculture of a country like Poland, which has just decided to move from a communist to a free market economy; whose farmers discovered that a free market is not such an easy, well-paying, and attractive alternative, and that it is not at all easy to export because the world market is very regulated by artificial bodies?

Many of you are already familiar with Poland. Companies like Pepsi-Cola, British Sugar, Central Soya, Heinz, and Campbell's Soup are actively investing or looking for investment in the country. Others, like Borden, Colgate, Kraft, Kellogg, Philip Morris, Nabisco, and Lipton have been heavily involved in trade with Poland for some time.

The problem is, we have an agriculture that employs about 25% of the active population. By the way, all over Eastern Europe, with the exception of Czechoslovakia, agriculture employs 20% or more of the population. Agriculture that is characterized by very low real wages and is, therefore, potentially very efficient and very competitive in international terms. Agriculture that is traditional and in many areas quite professional, but that lacks modern technology and equipment.

This agriculture has now discovered possibilities for increasing its productivity due to undistorted price signals, increased and improved availability of inputs and equipment, including imported ones, and increasing efficiency of agro-business, which is now being privatized in many areas, slowly, by foreign organizations. And artificial obstacles to private initiatives have been completely removed.

Now, however, this agriculture is faced by dramatically falling domestic demand, in per capita terms, due to decreased real incomes, change in relative prices, increased imports, cuts of waste by industry that has become cost-aware, households that no longer need to horde food, and the changing dietary habits of the population. And farmers cannot possibly count on being offered opportunities for industrial jobs, because industry is faced by 2 million people unemployed. This is the problem. Now let me give you some background.

Sometime in 1989 a national consensus was reached that there were two ways to solve the difficult Polish political and economic situation. One was a natural one: that thousands of angels would descend in Poland and would sort out the mess that we were in. The second was miraculous: that we stop discussing it and do something.

Poland has many reasons to believe in miracles, and we chose the miraculous way. Indeed, we started to reform our economy in a fairly rapid and quite radical way.

When the Solidarity government took power in September 1989, we were confronted with an economy that had been left in shambles by our predecessors. The country was approaching hyperinflation, some 2000% on an annual basis. The new government took drastic action: it liberalized most prices, cut subsidies, and introduced a comprehensive stabilization program. The Polish zloty was sharply devalued and made convertible for all current account transactions. Almost all fixed prices were freed, and those that remained controlled, such as electricity, coal, and gas, were raised sharply.

The government now has to act within a relatively balanced budget controlled by the parliament. Wages in the state-owned enterprises are strictly controlled. The central bank is pursuing positive real interest rate policies.

In two years since the start of this program, the change has been radical. Queues, the symbol of communism, have disappeared. Exports to hard currency markets increased by almost 50% in volume in 1990 and by a further 20% in nominal dollar terms in 1991. The year closed with a trade surplus. Inflation is down sharply to about 1–3% per month, still not a civilized level, but much less than 2000%.

Rapid progress is also being made in the transformation of our economy into one based on private enterprise and private property. Some 30–40% of the economy in value-added terms is in private hands. Private enterprise constitutes a major part of the wholesale and retail trade and an important part of road transport construction. One private firm employs 15,000 people and had a turnover of 300 million U.S. dollars last year. Half a million new private businesses were registered last year. Poland is definitely far in the lead among Central European countries in the development of the private sector.

Privatizing state enterprises has been a slower process. The first five enterprises were privatized through public offerings in December 1990. Trading of the shares began on the 15th of April, 1991, on the Warsaw Stock Exchange, which, by the way, is situated in the building that used to be the Communist Party Headquarters. The program for 1991–1993 envisages privatization of about 50% of the state assets. By 1995, the ownership structure of our country should be compatible with that of Western Europe.

Another area of reform in Poland has been the opening of the Polish economy to international investment and trade. The new foreign investment law was recently approved by the parliament. It enables full transferability of capital and profits, and puts an end to the requirement that foreign investors obtain permission to establish themselves in Poland (with the exception of very few areas of activity).

Foreign investors will find in Poland an extensive industrial base, a solid infrastructure for investment, a skilled, relatively cheap and young labor force, a market of 38 million consumers, and the neighborhood of the vast Soviet market. Safety of investment is assured by agreements on the promotion and protection of foreign investment that have been concluded with the majority of OECD countries, including the United States.

As regards trade, the economy is almost entirely open. We have the only properly convertible currency in Central Europe, and our average tariff rate is about 15%, with agriculture at 30%.

Except for such products as radioactive materials and military equipment, no licenses are required. With the exception of alcohol, no import quotas are applied. Producers are free to select their export and import channels. Open economic policy is considered essential for further progress in the economic transformation.

Change is similarly rapid and extensive in the banking system. Until now, the financial system of Poland was modeled after the communist prototype of the USSR. Despite a few specialized banks, it was effectively a mono-banking system dominated by the National Bank of Poland. While the banking reform began in 1982, the Solidarity-led government gave it a real boost. The NBP has acquired all of the authority normally invested in a central bank, notably for monetary policy, bank supervision, and foreign reserves management.

The system of commercial banks is constituted by some 50 medium- and large-size banks and a number of smaller entities. This network is still very small. The number of potential customers per bank branch in Poland is four to five times larger than in the United States. Although technically weak and not competitive, the network is changing and developing quickly.

These are only the most important areas of change. The parliament has enacted about 60 major laws on, among others, banking, foreign exchange customs, foreign investment property rights, competition, and privatization. A completely new tax system based on the value-added principle will be introduced in 1992. A uniform linear corporate income tax is already in force. A modern personal income tax will be implemented in 1992.

Now all of that did not come without cost, of course. The cost was high; indeed, very high. Unemployment, which Poland had not experienced since the war, became a fact of life. Production decreased dramatically with the elimination of subsidies, but also because of unfavorable external developments—in particular, the collapse of exports to the Soviet Union. For example, in only one year, exports to the Comintern countries declined by one-third, from $3 billion to $2 billion, and exports to the Soviet Union fell from $2 billion to $500 million. In addition, they now are unable to pay for previous shipments. As a result, healthy industrial plants that were producing mostly for the Soviet market are now almost bankrupt.

The decline in production has also created difficult budget problems. Large cuts in expenditures were necessary, and at the same time, the budget deficit increased.

The continuing decline of production, real wages, and public expenditures are heightening public discontent and frustration. These were reflected in the vote in parliamentary elections in late 1991. Some 40% of eligible voters decided to vote; 60% stayed home. The largest party collected only 14% of the total vote. The other large ones gained, roughly, between 8 and 11%. The communist party came in second with some 12% of the vote.

In a recent public poll, 32% of the population stated that it would prefer to live again under communism, and 30% said it did not believe the future would be better under a democratic system and a free market economy. Many of those who said so, and many of those who stayed home, were farmers: the turnout was particularly low in rural areas.

Election results pose yet another formidable challenge for the liberal reform-minded politicians, who despite the dramatic drop in public support, continue to be the majority. We need to take a new hard look at the social policy, with an aim of providing people with more security and certainty. We also need urgently to improve management of the state-owned enterprises so as to arrest a decline in production and budgetary revenue. Finally, we need to improve and enforce market regulations, such as competition rules, protection of consumer rules, and industrial standards, so the market economy is not perceived like the lawless Wild West.

If we do not act quickly and convincingly, the fundamental achievements of reform may be endangered. I have in mind particularly the open character of the economy, undistorted free prices, and a tight monetary policy. This is the internal side.

Now, on the external side, as I have said, developments were particularly adverse, including the collapse of the Soviet market and a certain collapse of our hopes associated with the EC. As you are perhaps aware, the EC has offered to conclude an association agreement with Poland, Czechoslovakia, and Hungary. We believe in Poland that this could become a blueprint for our reform: an association agreement that would be an adjustment program toward future EC membership. This would encourage other reforms, since no political party would propose and gain support for something that could delay our entry into the EC.

Unfortunately, the discussions with the EC took a different course. The agreement that is now almost concluded would be yet another trade agreement, with a considerable degree of liberalization, introduced into the relations between Poland and the EC. It would not have the character of leading to future membership, and setting related thresholds and requirements.

The important agreement in the agricultural field will, however, provide for considerable liberalization in the form of a managed controlled access to the EC market. For most traditional exports, conditions will be created allowing Polish produce to compete on the EC market, with the volume allowed to increase 10% per annum. This relates also to processed food.

We are now at the stage where we negotiate what 10% means. Does it mean linear or compound rate of growth? What produce? Is blood sausage included or not included? These negotiations are characteristic of any dealings with EC bureaucracy. When I think about our negotiations with the EC, the entertainer Tiny Tim comes to mind. One of his songs started with this little scene. Somebody was knocking at the door, and he asked "who is that?" "I am a viper." So Tiny Tim is frightened and does not want to open the door. Finally the caller says, "I am a viper; I came to vipe your vindows." We have definite difficulty in communicating with the EC.

DISCUSSION CHAPTER 8:

MR. MURPHY: One of the things that ran through my mind as you were talking—as Poland goes, so goes the rest of Eastern Europe. I wonder, as you think through your problems, what is the interchange going on with some of the other Eastern European countries, and how does that affect the thinking process relative to the EC? It seems to me it is not Poland alone; it is really a massive change that is going on in that part of the world. Would you care to comment on that?

DR. OLECHOWSKI: We do consult, we do coordinate our positions on some of the items. But we have conflicting interests, and we have very different systems. Let me use the following examples. Agriculture in Poland is a priority item for negotiations with the EC; it is not for Czechoslovakia. Poland has an open trade system, which is not the case for Hungary, which protects its agriculture by both high tariffs and quotas. So there is not much coordination or interchange that can take place in this area. Also, prospects for common trade are not that good with the different degrees of liberalism in the trade policies of these countries.

DR. VAN DER HEIJDEN: As you are aware, there is more freedom in trade between Poland and western countries in Europe. But recently there was a complete reluctance to accept importation of cattle or agricultural products from you and other Eastern European countries, and it was obviously just to protect the Western European farmers and countries. What is the approach of your country to that?

DR. OLECHOWSKI: You said, sir, that there was reluctance to import from Poland, but there has been a change for the better. There is, indeed, an important change in the EC approach, which says something good about politicians' boldness and farsightedness in this respect. The EC politicians made the decision to allow imports from Poland, Czechoslovakia, and Hungary to come into the EC at this limited but still fairly reasonable rate of 10%.

We are unable, however, to engage in a dialogue on where we go from here. How do we integrate these two important sectors sometime in the future? What would be the strategy and modalities for such an integration?

If we were to create one country in some future—and I believe we should and we will, say, 10 years from now—it cannot be one country with two agricultures. They have to integrate somehow. We do not yet have any consensus on that, or any meaningful discussion. This is what Waleska asks for often.

Now the second thing that Waleska asks is relative to our concern that given the change in the Soviet Union and in other countries in Europe, the EC subsidies may displace us from our traditional third markets—our traditional export markets. This is, again, the issue on which we are unable to have any dialogue with the EC.

DR. JACKSON: One of the reasons why there are difficulties in establishing the dialogue between the Community and the countries in Eastern Europe is that the Community is busy pursuing its own agenda of what is called "deepening" as opposed to "widening"—the discussions on economic and monetary union and on European political union, which are due to be concluded at the end of this year. Following that there will be a discussion of widening, first in the context of Austria, Sweden, and so forth, and then at the same time the perspective of the former communist countries in Eastern Europe.

I wondered whether the speaker could say something about how he sees this debate about widening and deepening. Do you think that the Community is pressing ahead too fast on the deepening agenda and that the commitments that are beginning to emerge—for example, the recent Dutch text on economic monetary union—will make it impossible for Poland to join the community in any realistic time scale?

DR. OLECHOWSKI: I do not see any conflict between our desire to join the community and a "deepening" versus "widening" debate. I believe that the aspiring countries like Poland should be told that they can join when they are ready, when they are ready to adopt these regulations, the regulations and agreements that will change all the time. That would be perhaps the best approach. The more ambitious, the more hardworking would be ready to do it earlier, while others may not be ready to do it at any time in the future.

I do not think that even an ambitious deepening would pose an insurmountable difficulty for Poland. To some extent, because of the nature of reform now and because we are changing our system completely, it is easier for us to accept institutions that may be difficult to accept by countries that have more entrenched systems.

One more debate that I think is also relevant here is the debate on common agricultural policy (CAP) in the EC. We obviously understand the difficulty that the Community is now under—the internal budgetary pressure to reform CAP, as well as the external (basically American and CAP groups) pressure to reform it.

We think that a third dimension should be considered. There are a number of Eastern European countries that should join at some stage; and, therefore, your CAP should be reformed in such a way that it would allow for acceptance of these countries in the foreseeable future.

MR. VAN ZWANENBERG: We hear a lot about the weaknesses of agriculture in Eastern Europe, and included in that is Poland. Could you tell us some of the strengths of the agriculture in Poland? What is it that we are going to buy from you? I remember historically Poland was a great country for bacon, black currants, and geese. Is it still? What are the products that we in Europe should be considering purchasing from you that are going to make some difference to a fairly full availability of food in Western Europe already?

DR. OLECHOWSKI: From the beginning of our negotiations with the EC we have emphasized the need to get improved access to the EC market for three sectors: fruit, vegetables, and animal sectors. We believe that we have comparative advantage in these sectors.

First, they are labor-intensive products and we have cheap and abundant labor. Second, they are not critically dependent on farm size; our farms are small and segmented, the average being between five and six hectares.

At the same time—and I continue to believe this is the right approach—we would slowly give grains, dairy products, in terms of processed foods, cheese, and so on. These are not the products in which we will be able to be competitive internationally. Therefore, we will be willing to give them. But we have to have major openings.

MR. MacMURRAY: H.J. Heinz is one of the companies you mentioned that has been looking around Central Europe for the last couple of years, in particular Poland, Czechoslovakia, and Hungary, to find that many of the companies we come in contact with have only the burning desire to export.

We have tried to explain that they can help themselves greatly by investing in their own future. The biggest business for most countries is the food business. If you want to change from 25% of your population subsisting on agriculture to a large number of people working in a value-added products business, then I think you should be looking to see how you can build successful food businesses within Poland to feed the Polish population—not on raw materials but value-added convenience foods.

We spent some time with one company, with the farmers and with the unions, and persuaded everyone to their satisfaction that it was to everyone's benefit to operate in the way that we would like; that is, by investing in the local farmers to help them to provide prepared raw materials value-added products that we would utilize in a more satisfactory way in the factory.

As I say, we convinced the unions, we convinced the farmers, we convinced the workers and the management of the factory; unfortunately, we did not convince the ministers in Warsaw. Perhaps you would like to comment on that.

DR. OLECHOWSKI: You obviously chose the wrong ministers. I am not sure you are providing the answer. My problem is that Poland is now a surplus agricultural producer facing falling—perhaps now stagnating—demand. It is not a question of Polish farmers being able to feed the Polish population; that is beyond doubt. Improvement in production techniques and increasing the yields would aggravate the problem further. We would have even more surplus and no ways of disposing it. This is our problem.

This is the problem also certainly of Hungary. And it will emerge very clearly in Bulgaria and Romania soon. I do not want to paint a doomsday scenario, but this will be quite soon the problem of the Ukraine and some of the former Soviet republics that may almost overnight become surplus producers.

We have been importing food for some two decades, with certain improvements in agriculture in the last three years or so and also with favorable climatic conditions in those three years. But because of the change in the economic system, we have turned into a surplus producer.

MR. VAUT: Poland has experienced the paradox of a food surplus and too many farmers that is perhaps difficult for a lot of people in the West to understand.

One of the first things to happen at the beginning of economic reform was that the infrastructure between food production and the consumer collapsed, including distribution. Along with that the food industry stagnated. Therefore, at least as a near-term solution, export of agricultural produce—live animals or frozen chilled carcasses, fresh produce and partially processed produce—is an important outlet for the farm community while somebody figures out how to get all of the rest of the infrastructure that you discussed—the Heinzes, the Gerbers, and the others of the world—organized.

It is very difficult. Those are low margin industries. It is hard to get a lot of new investment. There are a large number of factories of a very small scale.

So solving the problem of getting the domestic portion of the food industry going in Poland, as in Hungary, Romania, and Bulgaria, and, as will be the case, in the former Soviet republics, is a far more complex problem than trying to increase production. I think it is fair to look at Poland and the other countries and consider that, for them, agricultural trade in agricultural commodities in the short term is probably an important safety valve, if you will, to allow that portion of the economy to keep going.

It is important what Dr. Olechowski pointed out, that the farmers stayed home or voted recently against the Solidarity-supported reform program. Farmers took it on the chin as a result of these reform programs, as I think everyone predicted. That is going to happen in the rest of Eastern Europe and in the Russian Republics. There are no private farmers in the Russian Republics; but Poland has 25% of its people in farming, which increases to 40% of its people employed in the food chain when you add processing and the inputs industries. That is a big sector of the polity to simply put into a stagnated situation and expect the other 60% to keep the country going.

MR. URBAN: I am going to venture well outside my area of expertise. My experience is a take-off on the previous gentleman's comments. It is pretty much limited to Hungary and Romania. They also can feed their people, but they feed their people at a level where the value-added is so low that they are basically feeding commodities.

I thought the Heinz point was a good one, that there is an enormous opportunity to add value to products within the country that perhaps is being ignored due to the tremendous need to try to export commodities.

We in the United States feed our people—we could feed our people with maybe 5% of our production, but mostly what we feed is added value, not commodities.

When you repeated twice that you could feed your people, I thought I saw a curtain come down on the perception that added value within the country has a return. Would you want to comment on that, please?

DR. OLECHOWSKI: There is not much I can comment on. You are obviously right. There is the need to complete restructuring and reshaping of the food processing industry in Poland.

I did not mention this issue because it is complex and I did not want to use so much of your time, but to answer your question, while farmers, land, and farming operations are private, the food processing industry is state-controlled, cooperative-controlled, and extremely inefficient. Add to this the collapse of the distribution system, which is one of the phenomena that occurred in all Eastern European countries when the changing system was introduced. Now, no large companies could play an amortizing role in the market, develop the market and so on.

So there is a huge change needed there, the change that will not be done by us alone, that is for sure. It is quite obvious we need your expertise and your presence. This is one of the reasons why I am here.

But we do have this transitory problem of being unable to absorb the population. It took Western Europe 30 years to come down to figures of 7 or 8%. We believe we will have to do it in 10–15 years. It will be dramatic. It will pose huge problems, strains on social peace and so on, but we will have to do it.

DR. JACKSON: I was very interested by something Dr. Olechowski said: the importance of having a reform of the Common Agricultural Policy which not only took on board the pressures coming from within the GATT and the pressures coming from within the budget of the community, but also the dimension of Central and Eastern Europe.

What I want to ask him is what sort of CAP he thinks we could have that would take that dimension on board. Many of the features of the CAP that we do not like and that we are seeking to change within the community developed when the CAP was conceived against the background of a 15% share of the population in agriculture. That was one of the root causes of the type of system that we have.

We now have the proportion of agriculture down to on the order of 5%. That is one of the things that politically and demographically opens up the prospect of fundamental reform in the CAP. What kind of CAP could you have that would take on board countries which have 25%of the population on the land, and wouldn't this perhaps take us back to square one in terms of getting a decent rational agricultural policy in Europe?

DR. OLECHOWSKI: Well, the correct CAP would be no CAP and that is where we would like to get on board. We afford no assistance right now to our agriculture; that has to be remembered. We are looking now for opportunities for our unassisted and therefore particularly competitive and wild farmers to be able to enter the market.

What I have in mind is that 10 years from now, CAP will be different and Polish agriculture will be different. It will not be 25 % by that time, that is for sure. CAP will not be as ludicrous as it is today. I do not have in mind a precise scenario; I think, though, something which is in conflict with some of my other beliefs is that we would need to do some production sharing and some market sharing; that is, Eastern Europe with Western Europe.

In some areas it is not so difficult, for example, fruits and vegetables. It is meat which creates a lot of difficulties because it is, in fact, a zero sum game. Here we are saying make us some room; we will make you some room in some other products. This is how I see the future of European agriculture. The point is that we have to realize it is our common problem. Therefore, we should try to address it jointly.

MR. SCHWAMMLEIN: When the Berlin wall came down in Germany, the East Germans bought West German goods faster than they could be produced. That was in December, nearly two years ago. There is now a significant backlash: people in East Germany are now looking for East German-made goods. So, during these past two years, companies that invested in East Germany to produce goods equal to the quality and appearance of Western goods have enabled East Germans to buy goods made in East Germany.

What is the situation in Poland? Do they desire Western goods? As the gentleman from Heinz said earlier, if you would invest in Poland's industry and replicate the quality and variety of Western goods, wouldn't that help pull agriculture and industry out of their stagnation?

DR. OLECHOWSKI: I cannot but agree with you. This similar phenomenon occurred in Poland. There was an initial huge demand for Western food, because it is better packed, it is different, and so on. There was some return to domestically produced goods because they are fresher and cheaper, of course. There is some increase in variety and improvement in the appearance and quality of domestically produced food, but it is not yet anywhere at the rate of what is happening in Eastern Germany.

CHAPTER 9

The United States' Role in the Privatization of Eastern Europe

Carol C. Adelman

I am excited about our program administering foreign aid to Central and Eastern Europe, and now humanitarian programs to the former Soviet Union. I shall comment a bit on what some of the other government agencies are doing in trade enhancement and other areas that are the long-term solutions to Central and Eastern Europe's problems. Foreign aid is not the long-term solution; it is a temporary transitional solution.

I can look only in amazement at how different the world is now from how it was in the late 1970s—when international agencies were pursuing and supporting status policies as though development could be divorced from the private flows of investment dollars, goods and services, and ideas. It was a time when many in US AID were concerned that working with the private sector was helping the greedy and not the needy. The focus of donor assistance then—particularly in agriculture—was on the production end with little involvement in the value-added end, that is, storage, processing, retail, and distribution.

The revolutions of 1989 were amazing. The most tangible result is that we are sitting here with the distinguished State Secretary of the Ministry of Foreign Economic Cooperation from a free Poland. There soon will be a shipment of 650,000 primary school textbooks that are leaving from Dulles to go to Albania because they have asked us to depoliticize their textbooks from the last 50 years.

I was in Albania at the end of July with Deputy Secretary of the Treasury John Robson. When we arrived, Mother Teresa was there holding two children in her arms. She was taking them to Italy for operations on their club feet. We were proud to tell her that we were going to be providing relief and assistance to Albania. Mother Teresa looked at us and said, "Provide jobs." So the world has certainly changed a lot, and it is nice to know that Mother Teresa, as she probably has always been, is on the side of open markets as well.

Our own Polish-American Enterprise Fund, which provides loans to small and medium-size entrepreneurs in Poland, is located in the former Communist Party Headquarters along with the offices of an American group called the International Executive Services Corps, a group of retired executives located in Connecticut. The most interesting aspect is that my bureau and other bureaus in AID are doing direct contracts, projects, and grants with multinational firms, several of which are represented in the audience today—Land O' Lakes, Pioneer Seed, and Coca-Cola. Our East European program is 99% work with the private sector. So the world has changed, and I am pleased and proud to say that I think the government institutions around it are changing as well.

The business we are in at AID is giving advice. Although it is a dangerous and tricky business, we somehow stay in it, and I am sure many of you are involved in it yourselves.

We discussed with Dr. Olechowski the issues of both trade and the internal problems, production problems, and economic problems of Poland. I would like to mention a little of what the U.S. Government is doing on the issue of trade and our assistance to help resolve some of the domestic economic issues.

The U.S. Government effort is broad; it includes agencies such as Treasury, the U.S. Trade Representative and the Export-Import Bank. One exciting program that best exemplifies what we are doing is the Trade Enhancement Initiative, announced by the President in March of 1991. The goal is to confront the problem of the steadily falling trade of Central and Eastern European countries with the Soviet Union and to help boost their trade throughout the world.

We are pushing for a reduction of Western barriers to imports from the Central and Eastern European countries. We have relaxed our steel quotas and have concluded textile agreements that are already helping increase imports dramatically throughout Central and Eastern Europe.

Also, we are pushing for a reduction of agriculture export subsidies by Western countries. We have, as many of you know, a major initiative with the GATT, which we have had for many years.

Further, we are encouraging the former Soviet Union, as best as we can amid the turmoil, to remove or reduce barriers to imports from Central and Eastern Europe. We are also providing technical assistance and advice to the Central and Eastern European countries to help them with their own export promotion activities, having our customs service work with them to train customs officers, and providing technical assistance in competition policy, open and fair government procurement systems, and in other areas like this, all of which are new in these countries.

The U.S. Department of Agriculture is working with the countries of the region to help increase exports in products that are not covered by quotas in this country.

On direct foreign aid, we are primarily providing technical assistance and training. We have established Enterprise Funds in Poland, Hungary, Czechoslovakia, and Bulgaria. The idea is that we provide U.S. dollars to a board of directors comprised of private sector business people. The Enterprise Funds make equity investments or loans in enterprises in Poland or Hungary, and we get the bureaucrats out of second-guessing the investment decisions. Basically we are able to turn this over to a fund and monitor it simply from a fiscal fiduciary standpoint.

We also have contributed $200 million to the Polish Stabilization Fund. Fortunately this fund has not had to have been drawn down because the zloty has remained stable. The fund stays in a bank account and the interest earned will go to the Polish Government. Ultimately we shall jointly program that money with the Polish Government.

We have provided money for privatization assistance as well. At the request of governments, state property agencies, or ministries of privatization, we have firms lined up that we selected through open competition which are able to send over experts in privatization of airlines, dairy farms, and so on. These experts from a wide range of fields are sent to help out all of the Central and East European countries. We even have had three or four people go into Albania already.

We let big contracts in management training and education to the U.S. university and corporate communities, and we now have a program of about $18 million that links up U.S. business schools with business schools in Eastern Europe. They have already started this fall with U.S. professors in business, accounting, and finance working in universities in Central and Eastern Europe in training, having exchanges, and helping them establish their own degree programs in business training.

We are also, as I mentioned, providing much assistance in the area of commercial law, competition policy, and antimonopoly legislation. We are working to help decrease barriers to business formation, getting business licenses, and generally looking at the whole process of doing business.

All of these things and many others that I have not mentioned help the general business climate and also help agriculture and agribusiness; but in addition to that, we have separate agriculture and agribusiness initiatives that we are undertaking. One of these consists of assistance to cooperatives. Land O' Lakes has clearly been a leader in establishing cooperatives throughout the world; we are doing a big project with them. We are also doing a project with the American Cooperative Development Institute and the National Cooperative Business Association.

Additionally, we have a program in which we are letting contracts to agribusiness firms. While we have not announced this yet, because we are still in the selection process, I was delighted to see the level and the sophistication of the firms that bid on this. This project will help identify joint ventures in agribusiness, provide some financing for feasibility studies, and help groups find financing for the actual investment.

I also want to mention the importance of assistance to democratic institutions. We get requests ranging from helping to set up libraries in parliaments, to training jurists in establishing concepts in the rule of law.

We provided $12 million in assistance leading up to the first Polish elections in 1990, everything from providing loudspeakers to opposition parties, to newsprint for opposition newspapers that could not get paper because the state-owned newsprint companies still owned the paper companies. We gave them hardware and software for elections and sent over election observation teams. We even helped train political parties in how to organize for an election. We did this fairly, providing equal amounts of assistance to all of the democratic parties running.

Now that these elections have taken place, it is no longer appropriate for us to be that directly involved in internal political systems. So our assistance now is geared more to the parliamentary and judicial arena and helping establish independent media.

Another area we are beginning to work on is how to help stimulate the philanthropic assistance that is essential to a market economy in terms of social safety net. I do not want to ignore the safety net issues; about 17% of our program goes into that. I think that we always knew that communism could not deliver the goods economically, but the dirty little secret—or maybe it was not such a secret to those of you who lived in it or knew it and experienced it—was that they could not deliver the goods socially either.

We are appalled and amazed at what we are finding in the health sector. This is the only region in the world where life expectancies had been declining, where our Centers for Disease Control test vaccines that are worthless. These countries have the same disease pattern that we have in terms of cancers and cerebral vascular disease and heart disease, but the deaths are coming at a much earlier age because they do not have emergency medical systems; they do not have the ability to diagnose the problem early enough or to do preventive therapy against it. People are dying at much younger ages from the same things that we die of at later ages.

Also, we are now seeing the effect of the reforms in increasing unemployment and rising prices, and we are beginning to see what this is doing in the political arenas. We shall have to focus a lot more in this area. The Department of Labor is providing assistance in areas such as unemployment compensation systems, and we are working on vocational programs to relocate and retrain people.

From just a handful of democracies, our values of liberty and freedom are now the values that are spreading throughout the world. I think that the challenge to us, not just in the government, but in the private sector as well, is how to translate these values into the material benefits and the prosperity that we all enjoy.

I hope we shall be working together in this tricky business of giving advice. I look forward to a good partnership with the private sector in this effort.

DISCUSSION CHAPTER 9:

DR. TRANT: In food aid in the developing world, the concept of some sort of triangular transaction has become at least workable from time to time, where the food aid is provided by one developing country, purchased by an industrialized one, and provided to a second developing country where it is needed.

I suppose the question arises that the need in the former Soviet Union now and for some time is the need for food and its delivery, perhaps more its delivery than the food itself. I am wondering if Estonia, Latvia, Lithuania, and Poland, all of which have something in common as capable food suppliers, could not be a useful component of the aid exercise.

DR. ADELMAN: The U.S. Government has been exploring this area, as have the Central and East European countries. We want to explore these ideas to the extent we have the resources to do it. Our own domestic agricultural interests make it difficult politically to buy food from other countries and ship it to a third country. I think this is something that we may do more on a one-shot basis in terms of emergency assistance.

I think we need to be concerned, too, about delaying the inevitable decisions that the former Soviet Union needs to make in terms of their own internal problems and trying to normalize trade, undertake market economics, and focus more on reducing trade barriers.

DR. OLECHOWSKI: I could only add that the EC is willing to direct some of this assistance to the former Soviet Union and to Eastern Europe—not only to Poland and Czechoslovakia, but also Romania, Bulgaria, Yugoslavia, and the three Baltic states if the embargo is lifted.

The EC is willing to spend up to 50% of the budget for assistance, which is considerable because now they are talking about 2 billion ECU. I understand that the United States has legal problems with such a concept because food, in fact, cannot be purchased anywhere else than in the United States.

DR. ADELMAN: I think it is a matter of precedent and policy—we can purchase food from other areas, but we have generally done it for emergency situations.

MR. HETTINGA: Dr. Adelman, I have heard that there is massive corruption in the new Commonwealth. If that is true from your perspective, would you comment on some of the problems it has caused your agency?

DR. ADELMAN: AID is not yet in the business of conducting technical cooperation programs with the Commonwealth. So I have not had any direct experience in this arena. We have shipped $15 million worth of medicines into the area, but we were very careful to have a private voluntary organization on the United States ship carrying the medicines, to monitor them, and to establish relationships with the different hospitals.

We have heard horror stories about some other situations. Clearly, any foreign aid program or any kind of donations that you are doing, whether it is corporate or public, should make monitoring a priority in the program.

MR. TAYLOR: I think you mentioned that the focus was on helping the private sector. It does seem to me that there is an enormous problem, though, when you are dealing with economic systems that are in transition and where historically so much of the food sector has been handled on a command basis through a centrally controlled system.

It is difficult just to turn around and say, well, that is no good anymore; it cannot work and there is a better way to do it. There are large adjustment problems. The theoretical imperative that everything should go into the private sector is fine in the long term, but I think in the medium term it creates some problems. Could you comment?

DR. ADELMAN: When I say private sector, most of the people doing the technical advice or doing the training are from the private sector. But that does not mean that we would not be working with a state-owned farm to help it privatize or working with a similar institution—Securities Exchange Commission organization or the Ministry of Finance, for example, if they are trying to set up a two-tiered banking system. We would work with these institutions to help them set up the policy, legal, and regulatory framework to create the right climate for private investment.

DR. OLECHOWSKI: You are absolutely correct. I think it is particularly acute when there is such a need for some revival in the economy. The growth will not come because of privatization; privatization brings, in fact, a temporary drop in output. The growth will need to come from current assets, not from investment. So there is an issue of better management of current assets that are in the state-owned enterprises. That is why I mentioned this as one of the three priorities that I believe the government, the next government, will have to tackle. Otherwise, the whole reform program may be in danger.

MR. DOWNHAM: I might share with this group experiences from two days of meetings that we had in Moscow two weeks ago. On the first day we sponsored and participated in discussions pertaining to the management of diabetes and how, in particular, good nutrition and exercise can help in managing associated problems and help prevent its occurrence. The second day pertained to food technology and to opportunities for formulating products with NutraSweet, among other types of ingredients.

There was great interest. Representatives of the former Soviet Union and presumably other Eastern European countries are very interested in value-added products. If somehow or other we can find a way to bring that value, it is our judgment that people in those countries will readily consume these types of products.

DR. ADELMAN: I agree. We get these types of requests not only from the former Soviet Union but from Central and Eastern Europe. Consequently, we established something that we call the American Business Initiative. It is a $20 million fund to do joint feasibility studies with U.S. companies, the sort of up front, upstream work before investments. We do not have the money in this program to finance any of the physical infrastructure or do any of the capital projects with this; but in terms of the feasibility study work, we can help.

We have also funded some trade associations to set up joint trade associations in Central and Eastern Europe. One of the areas is agribusiness. One of our grants was to the Food Processing, Manufacturing, and Supply Association to go over to Central and Eastern Europe and begin to set up a trade association there for making contacts and encouraging business and joint ventures.

PROFESSOR HIGGINSON: Dr. Adelman, regarding your comments on the health priorities in the former Soviet Union, I think you might be misinterpreted as implying that a great deal more high-tech medicine is immediately necessary.

In the past, from my experience, localized centers of high-tech existed, but there was a general rundown of the primary and secondary health care systems that form part of the social infrastructure. One looking at the problem should recommend priorities.

It is of interest that the major causes of death in the upper age brackets in the Commonwealth, such as obesity, hypertension, and associated diseases, as well as alcoholism, have become more prominent in the late1960s and 1970s. Thus, overeating may have had quite an impact. Educational and other procedures, as well as control over nutrition, may be more important without necessarily newer technological approaches being applied.

DR. ADELMAN: I agree. Our strategy with very limited funds forced us to set priorities, which is what we have to do all the time; and we are not in any way excluding or saying that the preventive side is not important.

Our strategy was to deal at first with those areas that were the highest cause of death, to try to have as immediate an impact as we could through things like emergency medical services, and diagnostic and equipment needs; and it is not all high tech. In many cases we are talking about very basic emergency room equipment here.

But we have not forgotten or excluded the prevention end and are doing some work in that. The focus was on the immediate curative, to try to have some quick impacts.

MR. VAUT: I want to comment on the food aid issue with respect to Central and Eastern Europe supplying food to the Commonwealth. Aside from the obvious logic of Poland having a surplus in agriculture and the Commonwealth having a deficit in its food supply, at least at the consumer level, there is another side, another logic to that argument that often is not discussed.

I have just spent three years learning how to move agricultural commodities and food products around Eastern Europe with some efficiency. Supplying shiploads of grain and other products in airplane loads into international airports and major harbors is perhaps not necessarily how the Soviet Union food system was organized to move its food supply. There is a 40-year history of movement of food products from Poland into the Soviet Union, a lot of experience and infrastructure in place that argues for a greater efficiency for those kinds of movements than have been tried with other means. I do think there is some reason to pay attention to that.

The question that I had for our speakers would be an effort to link the morning's first discussion with this one. Secretary Olechowski referred to the interests of Poland in joining the EC and the steps to be taken to rationalize Polish systems, to standardize them along European norms. Dr. Adelman discussed all of the effort that the U.S. Government was putting into assisting Eastern European governments to learn to regulate and legislate along American lines.

I wonder if there is any potential inconsistency that will grow out of Poland, Czechoslovakia, and Hungary's interest in standardizing on European norms with a focus on the U.S. bilateral assistance effort to help transfer American norms.

DR. OLECHOWSKI: I do not think there is such a danger. The assistance that is used, in terms of providing some managerial skills to Poland or helping to educate in various areas, usually relates to relatively basic things that are shared by the United States and the EC. I do not think we have that type of problem yet.

DR. ADELMAN: It seems like it should be a problem, but I think I agree it has not been. Even though we might be sending people over to talk to them about commercial law, we do not expect that they are going to be taking on U.S. commercial law. We know they will be using European commercial law codes. Similarly with stock exchanges, we are sending over Wall Street investment bankers and stockbrokers to give advice, but it is on a much more basic level, and the countries have been adopting other models for their stock exchanges too.

PROFESSOR DUNLOP: We started these sessions with the full recognition that the food chain is a global industry, a global process. Indeed, all of its stages are more and more global. It would appear to me that we face in Eastern Europe and in the Commonwealth of Independent States particularly, as in some of the underdeveloped world, the question of how to add to the value within those countries and to find their place in the international arrangement. That is the problem. We will, I am sure, need a lot of discourse over a long number of years to work it out.

CHAPTER 10

Free Trade Between Mexico and the United States

Hermann von Bertrab

Mexico, the United States, and Canada are in the process of agreeing on a free trade zone, by means of which the whole region will achieve a greater competitiveness in the world. Let me tell you first about Mexico's position and then what we should be doing in the process of the Free Trade Agreement negotiations.

In the past 50 years or so, Mexico has industrialized under the concept of the infant industry argument that we all learned in school: You have to protect and shelter your newborn industry for it not to wither away with the bitter winds of competition. By doing so, it created a highly protected economy, highly overregulated, and one with a high degree of direct government intervention.

The Mexican Government obligingly started putting a wall around all new products being produced in the country, and we finally got to a system with 100% import licenses, that is, every item imported in the country needed an import license with tariffs up to 100%. This made a good shelter for new industry, but at the same time it produced a very inefficient system, because you do not have to be really cost-effective when you do not have to be competitive with the outside.

I did my Ph.D. dissertation on the transfer of technology from Europe to Mexico. When I did the research, I went to different industries in Europe that had manufacturing facilities in Mexico and said, what is it you produce here that you also produce in Mexico? You have Product B here and Product B in Mexico. What is the difference in your production process? They said, Mexico is such a protected economy that what we do is go into our files for the technology that is adapted to the market size of Mexico, whatever the present market size in Europe. They usually picked one of the outdated technologies and brought it into Mexico.

Thus, this system of inefficiency and high protection bred further inefficiencies because you had the wrong type of technology, or it was the right type of technology for that situation. But what you really had was a protected infant industry all the way to senility, without passing through maturity.

Further, it was a country and economy that was greatly overregulated, a maze of regulations. Except for the East, and India perhaps, we would have had a gold medal on regulations, but the wrong type. These regulations also meant less efficiency.

There was a high degree of direct government intervention into industry. Not out of a concept, like in the Eastern countries, but out of the sheer inefficiency of the system, and because the government took over the function of being the leader in development. It took over many industries that collapsed out of the very inefficiency and overregulation of the system. It was "Chapter 11," Mexican style. If you went bankrupt, you could not go on anymore, the government would simply take over and buy you out. The government got involved in sugar mills, textile mills, the steel industry, the telephone, banking—everything possible. Obviously things did not go better with the involvement of the government.

The government needed resources. So they just printed money, and created an inflation of close to 100%. On the other hand, we also borrowed money and there were some bankers launched with Arab dollars which—lending to a petroleum-owning country—piled up a loan balance of over $100 billion.

We were saddled with high inflation and a totally inefficient system. We had to pay 6% of our Gross National Product to finance charges on our debt. We needed a total refurbishing and restructuring of the country's economy, which has been done in the past years starting with President de la Madrid and followed by President Salinas with still more push.

Where are we now? The deficit has been cut from 17% to 2%—three times the effort that would be needed in the United States to balance the American budget. Tariffs and import licenses were cut. From 100 import items needing import licenses, we are now at three. Tariffs have all come down to 10% on average. The country has been unleashed from this inefficient system and you can see the energizing effect.

Let me turn to the Free Trade Agreement, the natural next step in the modernization of Mexico. It is the way to lock in and to further increase the modernization process, and makes sense because it would create a larger market.

With the same resources, the system that was very inefficient can now produce far more and have a market increasing for a long time. With the young population, there is an engine of growth for the whole region. There is everything needed to be efficient and competitive: financial resources, technology, less-trained people with low wages, highly trained people commanding high wages, and natural resources of all kinds.

Production processes can be adjusted in many ways. You can produce in Mexico what is better produced there, and in the United States and Canada what is better produced there. When people ask me to say what the Free Trade Agreement is about, I reply, everybody doing that at which he is best. We are in the process now of designing this Free Trade Agreement. It is like a game. It is a cooperation game within a framework of the competition of the markets. Involved will be goods, services, investment, intellectual property, and dispute settlement. We have to discuss rules of origin. Mexico does not want to be a pass-through; we do not want to take things from wherever and just pass them through into the American market. We have no interest in that. But if these rules of origin are too strict, you cut yourself off from the best possible sourcing, competitive sourcing outside, and you become inflexible and non-competitive.

You make your own people strong with subsidies, but we have to minimize or equalize subsidies. And health safety standards cannot be used for trade protection.

A well-designed agreement will minimize the disputes. Nevertheless, we shall have some problems, some disputes. So we have to devise a way to resolve them. Political things should be resolved more by consensus than by binding judicial decisions.

Finally, we must not push a country into something it does not like or have transition periods that are too short. These are all the problems that we are considering in order to design the best type of a Free Trade Agreement for the benefit of our three countries.

The commercial links of Mexico with the Eastern States are very old. The first American dollar was issued under decree of the Continental Congress in 1776. That first dollar was backed by the Mexican peso. I shall not say that in the future we shall back American dollars again, but we shall certainly add value not only to the American dollar, but to the whole region and to all of our lives.

The first stage of the negotiations is over, that is, exchanging information, understanding each other's problems, and trying to focus on the sticking points. That is where we are at the end of 1991. You do not resolve one question and then put it in a drawer and move to another question; you have to resolve the whole thing together. Nothing is agreed upon until everything is agreed upon. We are now in that process.

There was recently a meeting among the Canadian, American, and Mexican ministers. We started drafting the agreement and the agreement should be ready soon. The general system of preferences will be locked into the agreement so the United States concedes that to Mexico as a starting point.

We have many problems still to solve. From the point of view of Mexico, we must get a good substantive agreement; it is really no use having a watered-down one. Thus, we may have some problems coming, but we have made some progress.

Editors' note: On August 12, 1992 President Bush announced to the American people the successful agreement among Mexico, Canada, and the United States of a NAFTA treaty. The treaty must be ratified by the legislative bodies of each country. If ratified it will constitute the world's largest free trade bloc.

CHAPTER 11

Debt-for-Science Swaps: Mexico and the United States

The Honorable George E. Brown

The beginnings of the negotiations for a free trade agreement with Mexico are, in my opinion, extremely encouraging. I agree it is inevitable that we should move in this direction as a means of achieving comparative economic advantage, considering that we are going to be competing with other free trade areas. The European Community, of course, seeks just as we do to maximize the economic advantages of cooperation amongst the various nations of Europe, and they will be expanding to include the other countries of Europe, I am sure, even though it may appear difficult during this time of transition.

Some of the largest trading areas in the world are solely within the jurisdiction of one country. I think of China, for example. If you do not think that China is going to be a formidable trading bloc when it industrializes, you are mistaken; it will be.

I see the process of the North American Free Trade Agreement (NAFTA) among the countries of North America as being something that needs to take place in the very near future, then including Central America and expanding to Latin America as well. I think we are going to have a Western Hemisphere trading bloc that has a great deal of comparative advantage.

I am not an expert on this; I am not an economist. I am a politician, and I think this is going to happen because I think it is what the politics of the times demand.

Most of my time in Congress is spent dealing with issues of science policy. Of course in science we have already gone way beyond this sort of thing. Science is global today. America got its start by importing European scientists and European technology and acquiring the solid base we needed to become a great industrial power.

One might argue that a country at the stage of development of Mexico or the other countries of Latin America does not require a robust scientific establishment. I would agree that it is not indispensable. Japan demonstrated they could become a great industrial power without great scientific achievements. We did ourselves, in the period prior to World War II. But we soon recognized that we could not continue that way, and Japan recognizes that.

For a country at a stage of industrial development such as Mexico and most of the countries of Latin America, you do not need the world's finest scientific establishment to acquire the knowledge base of the world. What you need is a kind of transmission belt for technology transfer that will allow you to build the industrial strength and competitiveness that then will give you the luxury of supporting fundamental research in those areas that are attractive to you as a country.

That has been the focus of my efforts in the Congress: to achieve some understanding of the contributions of science and technology and to see how we can build closer relationships to a global scientific community and make it work more effectively.

I want to explain a proposal that has evolved over the last two or three years to expand our scientific cooperation with Mexico. It complements precisely the work going on with regard to NAFTA, which is aimed to provide Mexico access to a broader market, to new sources of capital, to various other things that you can get through such an agreement.

An effective arrangement for closer scientific cooperation achieves the same sort of results. It allows for the scientific communities of both countries to work synergistically on problems of mutual concern and to enhance the opportunities for economic development in each. So I have felt that perhaps we had reached a fortunate time in which things were going to come together that would benefit us both from the economic and the scientific standpoints.

Let me describe briefly the initiatives that I am talking about. We would like to link a reduction in Mexico's external debt burden directly to Mexican financial support for joint science and technology projects. Such a debt-for-science swap would operate the same way as the previous successful debt-for-equity or debt-for-nature swaps that have been worked out.

Discounted Mexican commercial bank debt could be purchased on the secondary market and retired in return for a contribution by the Mexican Government to either a specific research project or to an endowment. We favor an endowment that would fund a variety of research projects. This type of arrangement provides a mechanism and an incentive for a true binational partnership. Such a mechanism has been lacking in the past.

Let me make clear that we already have science and technology agreements with Mexico. We have had them for years, for generations; and they have been the most ineffective mechanism I have ever seen. They are signed with great aplomb by the Presidents of the two nations to celebrate their desire to work together, and promptly forgotten, at least in terms of funding. That was true of the 1972 agreement, which is still active, and it has been true of other agreements.

What we need more than a public relations agreement signing is a firm commitment of resources that can be used to support specific peer-reviewed scientific projects in which the two nations can cooperate.

The second initiative, closely related to the mechanism of debt swap, calls for the creation of the United States-Mexico Foundation for Science. On September 30 of this year leaders of the United States and Mexican scientific communities, including the two presidential science advisors, met in Washington to take the steps toward establishing such a foundation. Appropriate legal actions were also taken to incorporate the foundation as a nongovernment, not-for-profit organization dedicated to scientific and technical cooperation.

The general goal of the foundation is to provide a source of funding for high-priority research that will be carried out jointly by U.S. and Mexican scientists. The research may encompass a wide range of problems that face both nations, including agriculture, food safety, trans-boundary air and water pollution, energy, biotechnology, human health problems, natural hazard reduction, and science education. We certainly share problems in all of these areas.

Scientific priorities and policies for the foundation will be determined by a board of governors composed of the members of the scientific and technological communities of both nations. Research grants will be awarded through a peer-review process that will determine scientific merit, the degree of collaboration, and the potential for mutual benefit.

In order to provide start-up funding for this new foundation, Congress passed legislation that would provide several million dollars for a debt-for-science swap with Mexico. The State Department has not yet allocated these funds, but I am hopeful that they will do so soon.

The Mexican Government has already agreed that it will make a contribution to the foundation equal to the full face value of any debt retired for this purpose. Non-governmental matching funds also will be required and sought, and we have strong indications that both private industry and philanthropic organizations—some of our larger foundations—will be interested in contributing. I would be remiss if I did not use this opportunity to encourage all of you who have access to large amounts of money to make a contribution to this foundation when it is established. I think it would be of great benefit to your companies as well as to the scientific advancement of both of our countries.

It is my hope that the establishment of the United States-Mexico Foundation for Science and the successful transaction of a major debt-for-science swap will become models for scientific and technological cooperation between the United States and other developing nations of the world.

These nations must be our partners as we create a new world for the 21st century. I think it entirely appropriate that we begin this effort with our neighbors in Mexico, and it gives me a great thrill actually to be able to participate in bringing this about.

Finally, let me say that the idea for this foundation and the debt-for-science swap did not originate with me; I wish I could claim it did. It originated with scientists, including Mexican scientists teaching and researching in the United States at the University of California, the University of Texas, and one or two other institutions in the Southwest. They are well aware of the need for greater opportunities for Mexican scientists to do research. From their own experience as Mexican researchers working in the United States, they saw the importance of cooperation.

The charter of the foundation allows the funding of projects to be determined by a board of scientists. To the degree that they find attractive proposals for joint research in any basic science area, the charter will allow them to fund it. I rather suspect, of course, that a sense of priorities as to what will best serve the interests of both nations would require that at least part of the funding go to high-priority applied science research problems that will benefit both nations.

But I shall not pretend to spell out what the decisions of the scientific peer review board would be. The situation would be somewhat like that of the National Science Foundation board, which would prefer to spend all of their money on basic research, but under pressure from the Administration and from Congress fund applied research as well—and they are doing an excellent job of it.

CHAPTER 12

Changes in the European Common Agricultural Policy

Robert Jackson, M.P.

I shall discuss briefly the politics of farm reform and, specifically, the reform of Europe's common agricultural policy (CAP). Technological unemployment is something that concerns me as Employment Minister in Britain. Certainly, technological unemployment is a factor in the field of agriculture, in the sense that the fundamental pressure for change in European agriculture is coming from the development of technology. Science all over the developed world is making it possible to produce more and more with less and less labor. But the problems of agriculture are more than just another instance of technological unemployment, susceptible to the same kind of remedies that we use to deal with technological unemployment in different branches of industry.

It seems to me that something much more elemental is at stake in agriculture. The shift from the land to the town, from agriculture to industry or services, marks a fundamental change in the human way of life. And it is quite different, I believe, from that which is involved in the shift from a declining branch of industry to a rising branch of industry.

Let me illustrate this point simply and vividly. When you travel by train from Tokyo to Kyoto, the only beautiful things that you see, apart from the distant view of Mount Fuji, is the little pocket handkerchief Japanese farms. When I look at those closely cultivated, beautiful but highly inefficient units, and I compare them with the chaos and the jumble of the encroaching cities, I have no difficulty at all in understanding the Japanese desire to protect home agriculture, and that desire is driven by something more than simply the politics of the overrepresentation of rural districts.

In the longer settled countries of the developed world, at least, it is unlikely that agricultural production is going to be treated simply as yet another branch of production. There are always going to be powerful pressures to sustain farming as a way of life; notably, pressures to maintain the family farm, the cultivated environment, and the diversity of cultures. Despite that, the systems of agricultural support that prevail around the world, and notably in Europe, can't go on as they are. If we look at the European context, we have a system which is becoming unsustainably expensive to taxpayers, and increasingly inconsistent with the process of opening up worldwide trade in industrial products and in services to the GATT. But more than that, it increasingly fails to deliver satisfactorily the social and environmental objectives that alone can justify the system of agricultural support that we operate.

Let me put it briefly. The CAP is designed to support European agriculture by maintaining the price of European farm products. The result has been that a system that is intended to protect the family farm has advantaged the larger producer over the smaller producer, and a system that is intended to protect the land has turned out to be a massive engine for environmental degradation.

What do we do about it? We need to fundamentally redesign agricultural support, shifting the focus from sustaining overall prices to sustaining selected producers and selected patterns of cultivation. If we want to support a particular form of farm structure, for example, the family farm; if we want to secure specific environmental achievements or effects, such as the cultivation of the Alpine uplands or the wine and olive terraces of Tuscany; if we want to preserve regional diversity in foodstuffs—if we want all these things, then the best way to do so is to identify those social and environmental features that we want and pay specifically for them and, I might add, for nothing else.

I think this is the way in which the thinking about the European agricultural policy is developing, and I am pleased to see that. The Commission has made proposals that envisage substantial cuts in price support, something like a 40% cut in the price of cereals, which will be linked to the provision of direct income aids to different producers. There are signs on the political level that the Germans may be willing to adopt this approach, in spite of French opposition.

We shall need a considerable measure of understanding of all of this from the British government, because our farm structures in Britain are on average much larger than those on the Continent, and the consequence is likely to be that British farmers will suffer disproportionately from cuts in market support, which are compensated by income aids tied to the size of farms. That is something we have to recognize as being inevitable. And I think the answer for us in Britain is to emphasize the importance of aids that are linked to environmental objectives. We should also recognize that we shall gain if the overall budget for farm support in the Community can be reduced, and especially if we can achieve some shifting of it back to the national level and away from the Community level.

Understanding is also going to be required from the United States. I think that American negotiators in the GATT have done an impressive job in pushing the Community as far as they have, and that pressure needs to be continued and sustained. But alongside that, there is a need for continuing sensitivity to the factors that I have described, which make the politics of agricultural change so very difficult.

DISCUSSION CHAPTER 12:

PARTICIPANT: It was interesting to me to hear a European perspective. I thought I sensed a fundamental difference in the approach to regulation. Is the role of regulation to identify iniquitous processes, which need to be banned, or to identify positive things that should be licensed and permitted? I was curious as to whether there was really a difference in what may evolve as an EC approach, and the American process as you have seen it portrayed here.

MR. JACKSON: What strikes me is that the discussion in the United States on these issues seems to be much more transparent than it is in Europe. I think that probably has something to do with the nature of your political culture. Everything is put out on the table. You have a culture of litigation and an adversarial culture, derived in large part, I suppose, from Britain. The consequence of that is you get a very impressive disentangling of the threads of these issues; whereas in Europe it is rather typical with the political cultures, which have substantial prodemocratic elements in them, that these discussions take place more behind closed doors, are more discreetly handled, less confrontationally handled, less litigiously developed.

I think there is a difference of style, but I don't think that one would identify a difference in goals and objectives. In both countries one must recognize the extent to which impersonal values can be perverted into political ends. I thought that came out quite clearly in that issues of regulation can quickly become issues of protection. The European system might be more exposed to that than here in the States, but I am not sufficiently close to be able to really make that comparison.

DR. ARNTZEN: I want to make a point concerning your comment about shifting the basis for support for agriculture toward environmental issues. I strongly disagree with this approach, although it is also happening in the United States.

There is a shift in farming that is occurring in the United States and in Europe. There are more large farmers, and a smaller number of farmers are responsible for the majority of the production. In my particular state, however, in the last decade, we have seen an increase in the total number of farms, but when you study the location of new farms, they are largely around the major cities. These farms are a new form of recreation.

I don't think it is at all appropriate that we adjust or encourage government support systems for recreational farms. It doesn't add to our economic viability in the country at all. However, it is easier to castigate the subsidy supports for the large farms because they are easier to identify. The public gets more interested in attacking the rich guy who is getting support from the government. As we try to deal with this, we continue to shift our subsidy programs to a new basis. The shift in this country is to environmental issues.

I believe farm support systems coupled with environmental issues will be ineffective. Government policies work on a broad basis, but farming is done with a small, individual, location-specific set of conditions that don't lend themselves to broad government policies—especially in this diffuse area, the environment. What we are putting in place now is a long-term set of strategies that the government will implement with broad brush sweeps, and influencing individual farmers into the wrong decisions.

Lastly, if you are going to push in this direction, you are going to degrade the individual ownership of family farms. We see the new regulation on environmental issues coupled with more public regulation of the individual farmer. If he accepts a support for more grasslands across the fence rows, he is very soon also going to accept that the recreational land user from the city has the right to walk in those grass strips around his farm. He is gradually going to lose his individual rights on management of his resources.

There are specific examples where I believe the best use and the best environmental protection of land has been by individual land owners who have owned it for generations. They are concerned about the environment. They are concerned about maintaining their investment. What we have the potential to do, with the current push to include environmental regulations in farm subsidy programs, I believe, is erode individual land ownership, pride, and responsibility in the land. We risk replacing it with a government ethic telling farmers that the government knows more than they do about how this land should be managed.

MR. JACKSON: One of the arguments used to defend a system of supporting prices rather than producers is that you actually have a focus on the individual owner and there is a clarity about the situation of producing for the market. The problem that we are facing in Europe, and I don't want to comment on the situation in the United States, is that the cost of supporting agriculture by that route is becoming unsustainable. So, the choice you are beginning to face is whether you should have any agricultural support at all, or whether you should have an agricultural support that focuses on these sorts of devices.

The way to meet your very reasonable concern about how these things can be exploited is to look at the terms in which you define your environmental support and try to tie them down. We also have hobby farming developing in Europe, and I think that is a reasonable thing, but as you say, it shouldn't be supported by public funds. What you need to do is to limit your environmental support quite specifically. You need to tie it to particular regions of the country, where you can show that there are environmental factors that are relevant. I mentioned Alpine uplands, for example. That is a classic case. You need to look to the preservation of particular sites of scientific interest, areas of natural beauty that have been designated as such, where you want to preserve certain characteristics. You can, I think, tie down the kind of support that you are offering on environmental grounds in that way.

Of course, the political pressures will be to expand the area of support, to add to the catalog of possible environmental measures that can be invoked. We have to seek the politicians to balance that, to counter it. But I think the alternative for the farmer could well be either to accept some regime of this kind, in which he is being paid for outputs other than simply the production of food, or the loss of any support at all, and I think that would probably be a less acceptable alternative.

DR. ARNTZEN: Why did we get into this subsidy program in the first place? It went back to when we were trying to increase productivity in targeted areas of national need. We have gone way beyond that. By the mid-1970s, we had excess production capacity. We are now using subsidies to sustain a way of life. We are trying to maintain rural populations so they don't shift into urban centers. Why don't we recognize that it is a means of diversifying our economy by encouraging rural development, and that we would like to keep some of our population in rural areas, rather than shifting them into big cities that are already laden with problems? But don't try to disguise it with environmental issues and other things that are just going to have long-term negative connotations.

MR. JACKSON: I don't think we are really disagreeing, because what you are saying is "yes," if you want to support these people, support them, but support them on social grounds rather than environmental grounds. I am saying that I think there ought to be some environmental criteria as well. I am not sure that is a deep disagreement.

DR. MILLER: I agree with your statements about the two regulatory systems. We may approach it differently because of somewhat different backgrounds, but I think the goals are identical. They seem to come out the same hole most of the time.

The other issue, political union, would resolve many of the current controversies that are now ongoing in the area of food safety, for example, and in those areas in which national sovereignty is really a block. Is there any realistic anticipation that this will occur within our lifetime or the lifetime of our children?

MR. JACKSON: Political union is, if you like, a term of art. What has been going on in the Community over the last few years, since, I think, 1985, has been a sort of rolling equivalent of the Philadelphia Congress. And many of the same sorts of issues have arisen. What functions should be performed at the federal level; what functions should be left at the state level; how do we decide which functions these should be, and how should the decision-making be taken at these different levels?

The Community has proposals to expand the area of operation of the Community into new fields, and in those new fields to introduce more voting by majority in an explicit way, rather than having decisions made with a formal requirement of unanimity or continued discussion until everybody has agreed. Once you start to have decision-making by majority voting, the giving of greater powers to the federal political institutions, the European Commission, and particularly the European Parliament, is a democratic legitimate.

You ask specifically how that relates to the issue of food regulation. What is going to be required is establishing the extent to which this falls within the federal jurisdiction, as opposed to the regional jurisdiction. I think that the trend must be to put this onto a European level, basically because the fundamental principle of the Commission is the federal jurisdiction and what you call interstate commerce. So, there has to be some kind of interstate regulatory body simply for that reason.

The processes by which regulatory decisions are made in that field are likely to become more and more subject to majority voting, simply the process of harmonization of standards and so forth. Whether that will make decisions easier to reach, I don't know. In the sense that you will have one set of decisions relating to the 12 markets, rather than 12 separate decisions it should make it simpler.

But on the other hand, of course, the politics of operating supernationally, transnationally, multinationally, and multilinguistically are extremely complex and difficult. I often wonder how the Community can work at all. I think it is quite remarkable that it does. We are going to try and do it without a civil war.

CHAPTER 13

The Global Food System's Contribution to the Management of World Hunger

Gerald I. Trant

This chapter is a report card on the global food system's management of the world hunger problem. The global food system has made a great deal of progress in the last 40 years. Its achievements have been impressive in terms of production, but its grades are far from satisfactory when it comes to providing access to enough food of the right kinds to allow the world's populace an equal opportunity to live full, active, and productive lives.

What I propose to share with you this evening is a review of the global record of performance in two areas critical to ensure the livelihood and health of mankind: namely, that of food production on the one hand and food access on the other.

The production record of the global food system over the last 40 years is by any reckoning an impressive one. In 1951 there were only 2.5 billion people on the planet. Today there are more than 5.4 billion. At the beginning of the period and for a decade and a half after, despite the paucity of numbers—and that is by today's standards, not by theirs then—the food system was producing less than enough food for everyone. Things picked up, however; and from 1966 to the present the world has been producing enough food to meet the daily caloric requirements of everybody. In the developing world there has been enough food to meet caloric requirements just for the last 10 years.

This is a formidable record, brought about largely through yield increases, which have gone from 1.1 tons of grains equivalent per hectare in 1951 to about 2.3 per hectare presently. Perhaps it is wise to recall here that the cultivated land on which this food is grown amounts to only 1.4 billion hectares, that is, 3% of the surface of the world. It produces more than 90% of our food, about 93%, in fact. The other 7% comes from marine sources and from pasture.

Various factors have contributed to these production increases. Chemical fertilizer use has increased 400% or more over the period. The uses of pesticides and herbicides also has grown rapidly, along with better crop management practices. In the developed world, favorable price relationships—and that is probably an understatement in some areas—between products and factors of production have provided a significant stimulus to increased agricultural production. The growing use of petroleum-based powered farm machinery has enhanced the productivity of both land and labor. The list of credits certainly would not be complete without reference to the development of high-yielding varieties of wheat, rice, and maize that responded favorably to modern management practices and particularly well to high applications of nitrogen fertilizer.

Although it is true that production increases have exceeded population growth rates for the world as a whole and for the developing countries in particular in recent years, the challenges ahead are nonetheless daunting. There are good reasons for believing that in the next 30 years or less, we shall see a population increase of 60%. In more graphic terms, this will mean that Asia will have to feed a new China plus. In Africa we shall have to feed more than two and a half people for every person living there today.

The prospects for food security in sub-Saharan Africa are not encouraging, because although food production has outpaced population in the developing world taken as a whole, most of the countries of sub-Saharan Africa have seen food production per person decline for the last 20 years. Even the production prospects necessary to maintain food security in global terms give real cause for concern. There is no large stock of unused technology available now as there was in the beginning of the 1950s, and the high-yielding varieties of rice, for example, now seem to be reaching a production plateau of about 4 tons per hectare. Additionally, Africa, as you know, has yet to have its own green revolution. Difficult though they are, the production problems and prospects seem to be substantially less than those having to do with access to food. The sad and frustrating truth of the matter is that despite the food system's ability to produce enough food for everyone at present, despite the widely accepted importance attached to that ability, still not everyone has enough to eat.

There are many kinds of hunger and malnutrition. Probably the best known is famine, which affects everyone in a region. Even though it is temporary, it is devastating for the people affected. Time was when famines were most frequent in Asia, but recently they have been concentrated in Africa, where their causes are frequently a combination of natural disaster made worse by civil strife or war and indifference.

About 30 million people, most of them in Ethiopia, Mozambique, and the Sudan, face starvation as a result of drought and war. Serious food shortages are being faced in 14 other countries. Famine, because it is so frequent in Africa, has become a chronic crisis rather than an exceptional phenomenon. The same causes, war and bad weather, have caused serious food shortages in other regions as well.

Although famine is probably the best known type of hunger because it lends itself dramatically to media coverage, there is, beyond the suffering and starvation and death it causes, a much larger number of people who are chronically hungry or undernourished. Their numbers have crept up over the years, from 450 million in the early 1970s to 550 million today. Of that number, which put in perspective means about 11% of today's total world population, about 160 million are children five years old and under, who are suffering from protein energy malnutrition that will in all likelihood stunt their growth and development permanently. Forty thousand children die each day from ordinary and preventable malnutrition and disease.

Specific nutrient deficiencies afflict many millions of people. For example, between 20 and 40 million people suffer from a shortage of vitamin A and are consequently either at risk of losing their eyesight permanently or have done so already. There are 200 million people afflicted by endemic goiter caused by iodine deficiencies, as you all know, with an estimated 800 million more at risk. Iodine deficiency in pregnant women is the cause of cretinism in children. It is a life cast in misery for a child who will never develop, who will never have the mental capacity to do so, and it is due to a very simple cause. And iron deficiency anemia affects more than a billion and a half of the world's population, reducing their ability to function independently.

Food insecurity is not evenly distributed around the world, although it is present to some degree in all countries. As you would imagine, by far the greatest number of hungry people live in the developing world. About 80% of them are the poor who live in rural areas, although the number of hungry people in urban areas is increasing, particularly in Latin America. About 60% of the hungry people live in Asia; about 27% in Africa; 10% in Latin America; and about 5 percent in the Near East. While it can be seen that hunger is largest in Asia, its growth rate is fastest in Africa. But even 10% of the U.S. population is now depending heavily on food stamps.

The prediction for the next 10 years and beyond is not encouraging, particularly for sub-Saharan Africa. In eight years, the number of hungry people there is projected to grow by 25 million people to more than 165 million. Some estimates go up as high as 200 million hungry people. Assuming that food production and population trends continue, the gap between domestic supplies and requirements could well reach 50 million tons per year. Closing the gap would require more than doubling the present rate of growth in current food production levels of 1.7%. A target like that, as you will well recognize, is extremely difficult to achieve.

Present levels of food aid—I think there are about 11.4 million tons currently—are unlikely to be increased substantially. In any event, they are likely to be partly diverted for some time to the erstwhile centrally planned economies of Eastern Europe. While it is claimed that food shipments to Eastern Europe are additional to food aid to developing countries, there is cause for concern that they may serve to put a cap on shipments to the growing numbers of the world's food-deficit developing countries.

It is here perhaps pertinent to recall the additional complications posed by the uncertain prospect of the agricultural component of the ongoing multilateral trade negotiations. If agricultural price supports in the industrialized countries are reduced, as appears to be a remote possibility, production surplus to domestic demand may well be reduced with a concomitant reduction in food stocks available for food aid. The requirements for aid are neither going to reduce nor level off for the foreseeable future, although food problems in other regions of the world are likely to be less severe than those projected for sub-Saharan Africa.

The greatest number of hungry people will continue to be in Asia—specifically the rural areas of South Asia. It is expected that the proportion of the world's hungry that live there may decline as a consequence of improved economic circumstances and increased food production during the 1990s. However, the new estimates of Asian population growth from the United Nations, which puts it at 1.8% and not 1.5, suggest upon recalculation that the number of hungry people will not be reduced in a substantial way; perhaps by 40 million.

The food production situation in Latin America and the Caribbean is not encouraging. Production per person for the region as a whole appears to have stagnated for the last 10 years. As might be expected in such circumstances, production per person in some countries has been declining. I cite the countries of Peru, Bolivia, and Brazil as cases in point. They have been having difficulties increasing food production rapidly enough. In fact, the Incas were doing a better job in Peru than the current Peruvians in terms of food security.

Hunger and poverty are closely linked. A major investigation into the magnitude and evolution of poverty in Latin America and the Caribbean, recently completed for a UN conference, indicates that a third of all Latin Americans live in chronic poverty; a third, 33%. If they use a slightly less restrictive definition, 60% of the population could be swept in and considered poor. It is discouraging to note that the optimistic projection indicates a constant proportion but a growing number of poor in the region, reaching about 300 million by the year 2000.

It would be good if, after inundating you with facts that appear only to reinforce the negative, I could turn a fresh page and, in the words of the old song, accentuate the positive. I find I cannot. There are no quick and easy fixes for the world's hunger problem.

It is, however, encouraging to see that some recent progress has been made in conveying emergency food and medical aid across combat zones by means of relief corridors, or corridors of tranquility, as they have been called. Where food emergencies become chronic, as they have become in parts of sub-Saharan Africa and elsewhere, their disincentive effects on local food production have to be taken into account.

There is no question that only national governments have the resources to deal with the hunger problem of their people. Ensuring food security for a family on a permanent basis requires that income or resources are available for that family to buy or grow its own food. Economic growth, as measured by gross national product per person, makes food security more affordable. But let me assure you, it does not guarantee that the poor and hungry will be fed. Countries wishing to eliminate poverty, hunger, and malnutrition will have to establish their own programs to do so. In the developing world there are many examples of successful programs to combat hunger. China, for example, has an effective rural production and employment program that has reduced the proportion of hungry people to 6%, a record that many higher income countries have yet to achieve. Other effective programs are those in Sri Lanka, Chile, Cuba, and Uruguay.

Internationally, there are encouraging signs of action. The World Bank's recent orientation toward the elimination of poverty through labor-intensive growth and increased spending on social services has a great deal to commend it. Current estimates are that there are more than a billion poor people in the world, which in more manageable terms, means one in every five of us.

Recently the United Nations adopted in its international development strategy for the 1990s the four hunger alleviation goals enunciated by the World Food Council in its Cairo declaration of 1989. Those goals are as follows: (a) the elimination of starvation and death caused by famine; (b) a substantial reduction in malnutrition and mortality among young children; (c) a tangible reduction of chronic hunger; and (d) the elimination of major nutritional deficiency diseases.

There are other efforts being made to reduce hunger in the world. Bilateral programs between industrialized and developing countries are often effective because of their magnitude. Nongovernmental organizations are frequently effective in complementing the work of development agencies, both bilateral and multilateral. Another important contribution to world food security has been made by the international agricultural research centers of the consultative group on international agricultural research—financed by the World Bank countries and foundations.

Despite all these contributions to world food security that are in addition to those made by the global food sector that you represent, the objective of global food security remains elusive. In terms of its report card for managing the hunger problem, the global food system deserves a B minus. It could have been worse, you know; but I was always easy on my students. More people are being fed than ever before, but the same proportion is hungry.

On the technical side, the global food system has performed relatively well. Production has more than doubled in the last 40 years. There is the prospect of continued production increases in Asia. But the production problems of Africa have yet to be resolved, and concerns are developing for Latin America and the Caribbean. For its production achievements, the global food system deserves an A minus.

When it comes to assuring access to enough food for all the world's people, the global food system has made little progress. True, more people now are being fed than ever before, but more people are hungry than ever before. In terms of achieving access to food for all the world's people, a grade of C minus. Actually, I had a D earlier on. After all, shouldn't everyone have the right to enough food?

DISCUSSION CHAPTER 13:

MR. MURPHY: All over the world the United States is involved as a supplier of rice. The current production figures are caused by governments, and I say governments in plural, and certainly the United States amongst them. Caused by war, caused by using food as a political ploy, whether it be internal, as in Nigeria, or external, as with Iraq, where it is a punitive thing—to starve 20 million to get one guy out of a job. I would be interested in your comments on how we get around this use of food as a method, as an irrational way to accomplish local objectives.

DR. TRANT: It is an anathema to me that the innocent should suffer in the way that they do. To cut off food supplies or economic opportunities to a country is extremely difficult to justify in terms of any of the principles for which the United Nations stands. It is within the powers of the Congress of the United States and the Administration to provide the answer for that question for this country.

MR. HETTINGA: Dr. Trant, you painted a pretty grim picture about sub-Sahara and Central Africa. Would you comment on the fact that the current AIDS crisis represents 30–40% of the young population? How is that going to affect the food supply? You have not only food supply problems, but you have a whole population being wiped out.

DR. TRANT: It is hard to get the data on those heart-rending cases. I would make this observation: if you wipe out the productive-age population, which is what will be hit or is being hit by HIV and the AIDS syndrome, then the productivity of agriculture takes another dive, and there will be fewer people with less to eat. So the problem of hunger does not disappear.

I would only be speculating in trying to achieve some balance between the population growth rates of Kenya, which are in the vicinity of 4.2%, and those kinds of statistics.

PROFESSOR HIGGINSON: I understand that you are implying two things. First, that regions should be relatively self-supportive, on the grounds that food transportation systems, from a practical point of view, are not going to provide a solution, as they did in the 19th century and early 20th century between America and Europe, for example.

Second, you have implied that there are multiple approaches to this problem, but you have not mentioned "population dynamics" as a possible means of reducing the pressure on fragile governments. What are your views on the latter as part of an overall general food policy?

DR. TRANT: I make the observation that since about 80% of the world's people live in rural areas, and since transportation within them is not terribly good, even though you can move between regions, the chance of being able to get food to where it is needed except on an exceptional emergency basis is pretty low. The cost of moving food in on an emergency basis is enormous per ton of food delivered. In some cases, with an air drop, you are talking a cost factor of maybe 10 to deliver it.

PROFESSOR HIGGINSON: To be strictly honest, shouldn't we face up to the problem of overpopulation when we discuss food supply?

DR. TRANT: I think right now we already have a problem. The amount of food that it will be possible to produce on a continuing basis if you take, say, North America, European, and Australian levels of consumption, which are based on about 16,000 gross kilocalories per person, is no longer available for the current world population, and there is no prospect of that.

Population growth is, fortunately, now becoming more generally accepted in the developing countries as part of their problem. At the World Food Conference in Iowa a couple of years ago, a lady from Kenya said, well, you just do not understand it. We have to have large families because our death rates for children are so high and we depend on children as our social security system and do not bug us with this issue.

Now that the health situation in the developing world has improved immensely and they are recognizing this, the prospect of longevity in a developing country is now at least equal to what it was in northern North America 50 or 60 years ago; a really remarkable change.

So there is some recognition that there is a problem. The next step is doing something about it. That, I am afraid to say, is some distance in the future.

DR. GAULL: You gave the world food system rather mediocre grades for accomplishment. Yet, in the United Nations only a short time ago, multinational corporation was a four-letter word. Recently there has been a major political revolution suggesting that perhaps the political model of many of the developing countries may not be the ideal one.

Do you now see, with the new political situation, the possibility that the United Nations would be able to work with these behemoths that they eschewed just a few years ago?

DR. TRANT: The Center for Transnational Corporations was set up in the United Nations at least in part on the insistence of the Soviet Union, and the idea was to beat the international corporations over the head for all the terrible things that they were supposed to be doing. They were joined in this by a number of developing countries who felt that all of their problems lay elsewhere; they needed a new economic order and all the rest of that before anything could be done at home.

So there was nothing better than going after the transnational. The U.N. has a transnational center, and I think it would be interesting for a number of transnational corporations to approach said center at this juncture and say, look, we think that we could maybe offer assistance in some of the areas of poverty and employment where the developing world needs assistance, and see how they reacted. It could be, I suppose, an initial offer for discussion of the idea. But I think the idea would have some merit, and it would turn the whole center and its orientation around. Not that it has all been entirely negative, but it certainly was not set up, in my view, as a positive thing.

I think that would be worth taking a look at, particularly in the food sector.

PROFESSOR SCHELL: It is often said, and I think it can be argued convincingly, that the new technologies in food production that we discussed will have considerable potential, and could have major impact on the solution of problems of food production for the Third World.

From your perspective, is this realistic or wishful thinking? And if you would come to the conclusion that it is realistic, what would be the best way to implement it?

DR. TRANT: That is an interesting and complex question. The World Food Council did hold a consultation early this year in Cairo, and it was directed toward the need for another green revolution.

The general assessment was that Africa certainly needs something because they never did have anything in the first place. The revolution has hardly touched Latin America, and there are concerns about barriers to increased production being reached in Asia and elsewhere for the high yielding varieties.

In terms of biotechnology, I think my view is that it will assist in achieving more rapidly the kind of changes in the genotype that one would seek to attain, because it is a more efficient medium for genetic manipulation than we had before. Sort of like going from a light microscope to an electron microscope.

Biotech and bioengineering in and of itself does not tell you what the plant ought to look like or what should be done about it. You have to have that initial capability and a sense of direction. So the techniques, to the extent that they accelerate progress in the appropriate direction, would be most helpful.

I hope that it would be possible to work with the varieties of plants that are typically grown in, say, regions of Africa. I think of the cocoa yam as a case in point, and of cassava, chick peas, and the various sorghums. I hope that the rate of development could be fairly rapid there; but when I think of how the present day green revolution came about, I have to remember that there were maybe some 20,000 cultivares presented to IRRI by the Japanese government. This was a gift without equal. By using that germ plasm, it was possible to make substantial progress very rapidly. I do not know where such germ plasm stocks are for the food crops of Africa at present. I think more could be done.

MR. MARESCHI: My question deals with the safety aspects of food, and the role that food-borne disease has in the developing world: what kind of impact does food-borne disease have on the economic development of those countries?

DR. TRANT: I can only make some observations on that. Obviously, quality control is more difficult to achieve in the developing world than it is in the developed world.

The use of pesticides sometimes may be excessive beyond belief, but as far as getting an actual measurement of it, I do not have one. Typically the resources are not available to carry out reasonable supervision. The technical resources are not there. The money to pay the people is not there. I suppose when food is really short, it is a question of producing more, rather than being overly concerned with the quality of it. We cannot accept that as a desirable state of affairs, but it tends to be the way the world is.

PART IV

DIET AND CANCER: THE POLITICS OF RISK COMMUNICATION

Dorothy Nelkin

Over the past 10 years, the American public has been deluged with conflicting information about dietary risks—from the claims of corporate advertising, the recommendations of scientific panels, the warnings of consumer groups, and the reports of the media. We read that our choice of foods can encourage cancer, or can help prevent this dread disease. We hear that concerns about dietary risks are irrational, a manifestation of "cancerphobia"—or that cautious diets are a highly rational way to protect one's health. Scientists themselves convey conflicting conclusions about the safety and efficacy of different dietary regimens, and especially about the relationship between diet and cancer. As in so many other risk disputes—over industrial pollutants or consumer products—groups with different interests and ideologies profoundly disagree about the nature of the risks and, especially, the information that should be communicated through the media and advertising.

There are many stakes involved in risk communication, and most risk disputes involve prolonged and bitter struggles to control the information conveyed to the public. For it is widely assumed that media messages can change consumer behavior and convince the public and regulators about the dangers or the acceptability of specific products or policies. There is far more to the communication of risks than simply the disclosure of technical risk, for interpretations of risk are influenced by economic and personal stakes, professional ideologies, administrative responsibilities, and professional beliefs.

Communicating risks about cancer is an especially sensitive area. Consumers are alert to cancer warnings in light of the growing burden of this disease and its dread image. Whenever evidence emerges linking cancer to food—whether from pesticide residues or diet—consumers demand information, assuming that changes in lifestyle might enhance their personal control over disease. Yet communicating information about cancer risks is complicated by intrinsic scientific uncertainties, skepticism about the credibility of experts who are often viewed as advocates, and sensitivity about the influence of tentative research findings on consumer behavior.

This paper seeks to unravel these dilemmas by reviewing three specific disputes that took place over risk communication in three different arenas: the media, product advertising, and scientific guidelines and dietary recommendations. In each of these arenas, disputes developed over the timing of reports, the adequacy of evidence, and the extent of scientific consensus necessary to support a claim. Much of the debate in the cases I shall describe focused on technical questions. However, the cases also demonstrate that underlying technical disputes is a set of critical political and social issues—the proper role of government, the ability of consumers to make informed choices, and above all, the impact of risk information on a highly competitive industry. Because this paper is intended to generate discussion, I shall, in each case, pose the issues as questions to be considered by the panel.

DIETARY RISKS IN THE MEDIA

In March 1981 the *New England Journal of Medicine* published a preliminary epidemiological study suggesting that coffee drinking might increase the risk of contracting pancreatic cancer. Pancreatic cancer is among the more common forms of cancer in the United States, killing about 20,000 persons annually. And coffee is, of course, a major consumer beverage. Thus, the findings of the study, though tentative, were of considerable public interest—that is, regarded as newsworthy by the press.

The *NEJM* is an important source of information for the media, and the study received very widespread coverage. Popular articles in magazines and newspapers properly noted the methodological limitations of the study and the lack of conclusive proof. But they also insisted, as a *New York Times* editorial put it, that the study "ought not to be dismissed."

The study, and especially the publicity surrounding it, drew harsh criticism from scientists, and it raised questions about the role of the media in communicating tentative scientific results to the public. A commentary in the *Journal of the American Medical Association* observed that the coffee report was presented "in a pattern that has become distressingly familiar." It was given "widespread publicity before supporting evidence was available for appraisal by the scientific community, and the public received renewed fear and uncertainty about the cancerous hazards lurking in everyday life." Scientists criticized the methods of the study, but they also questioned the wisdom of informing the public about such risks. The *JAMA* comment stated, "Whether to shout 'fire' in a crowded theatre is a difficult decision even if a fire is clearly evident. The risk of harm seems especially likely if such shouts are raised when the evidence of a blaze is inconclusive." The writer worried about the effect on the credibility of science as well as the confusion of the public.

This critique illustrates recurrent dilemmas in the reporting of risk. When should scientists publish results that are uncertain, but have potentially important implications for public health? How should tentative findings about risk be communicated? Is the reporting of risk information like shouting "fire" in a crowded theatre or is failure to report potential risk simply "covering up" sensitive problems? News reports are often the only means through which the general public becomes informed about hazards. The selection and construction of news about risk help to shape consumer attitudes and to define public policies. Thus, the role of the media is at the center of disputes over risk communication. Although consumer advocates and journalists want immediate and total disclosure, some scientists believe that "the rush to the press is simply mindless, if not unethical ... until data are interpreted and validated ... and until all currently available data on the incidence of human cancers in exposed populations can be integrated." Taken literally, such constraints on releasing information would totally preclude public communication.

SCIENTIFIC RECOMMENDATIONS AND GUIDELINES

In 1980 the Food and Nutrition Board of The National Academy of Sciences published a short report, *Toward Healthful Diets*, contending that it was "scientifically unsound to make single, all-inclusive dietary recommendations" regarding cholesterol, fats, fiber, protein, and carbohydrates, and that there was no basis for recommending dietary changes to prevent cancer. The meat, dairy, and egg industries hailed the report. But some scientists, convinced that epidemiological, metabolic, and animal evidence called for the reduction of cholesterol and fat consumption, contended that the report applied unrealistically stringent standards of scientific proof to the data, and that failure to offer dietary advice was irresponsible. They accused the authors of personal and professional bias.

Two years later, a new NAS report, *Diet, Nutrition and Cancer*, proposed interim dietary guidelines that might reduce cancer risks, though it cautioned that the evidence was not conclusive. The recommendations included the reduction of fat intake to 30% of calories. This too was controversial, criticized because it was not adequately supported by evidence, and uncertainties remained. Once again, critics questioned the motivations of the expert panel. The American Meat Institute was especially critical of the news release issued by the NAS, for it had mentioned specific meat products such as hot dogs and ham. This they felt "represents an inexcusable distortion of a rather narrow set of facts." Government, they said, should not be telling people what to eat.

The disputes surrounding dietary recommendations involve questions of the interpretation of evidence and the timing and content of risk communication. Scientists with different notions of adequate evidence reach conflicting judgments about the strength of the link between diet and disease and the wisdom of making public recommendations. Each side in these debates accused the other of predetermined policy preferences and nonscientific motivations that biased its approach. A debate that began with technical issues spread to include criticism of the NAS procedures for selecting panels, choosing staff, weighing evidence, and reviewing reports.

ADVERTISING AND PROMOTIONAL CLAIMS

In the fall of 1984, The Kellogg Company began to promote All-Bran cereal by associating the product with cancer prevention. The ads exploited the National Cancer Institute recommendations for a low-fat diet, proclaiming, "At last some news about cancer you can live with." One ad pictured an All-Bran box, a bowl of cereal, and a bar graph comparing the fiber content of the cereal with several other foods. This was the first advertising campaign in which a major food manufacturer implied that its product may reduce the risk of cancer, and it sparked interest in an industry eager to attract health-conscious consumers. But it also sparked a debate involving the FDA, the FTC, the food industry, the National Cancer Institute, consumer groups, and several scientific and medical organizations—all concerned about the proper regulation of health claims in advertising and product labeling. Should such claims be allowed at all? If they are permitted, how should they be substantiated? How much scientific consensus should be required before health claims are made? As the California Cancer Advisory Council put it with some prescience: "If Kellogg continues and no one objects, then next week we could see broccoli or whole wheat sold not as a food but as a preventative for cancer."

The dispute was compounded by ambivalence over the role of government regulation, and debates among regulatory bodies. The FDA had long held reservations about the legitimacy of making health claims for food, for it was determined to maintain the integrity of the food label as a source of reliable consumer information. The agency stressed the pitfalls of health claims, especially in light of disagreement within the scientific community about diet-disease relationships. But other agencies supported the practice. The NCI, for example, saw such advertisements as a cost-free way of reaching millions of people with its message about dietary fiber. An FTC official endorsed the All-Bran campaign as communicating important public health recommendations to the public. When the FDA decided not to take regulatory action against Kellogg, other food manufacturers pressed for clarification of the situations in which health claims would be permissible.

The incident opened a long discussion about the way to regulate health claims. How can messages be designed that do not oversimplify the scientific uncertainties, yet are still appropriate for publication? What should the FDA do about claims that do not tell the "whole truth," for example, by promoting a product as low in cholesterol when it is also high in fat? And how can industry avoid starting nutrient "power races" in which advertising claims lead competitors to add more and more fiber to their products? In the days of so-called nutraceuticals, when food has become a kind of medicine and part of the anticancer arsenal, such questions will assume increased salience.

COMMENTS AND CONCLUSIONS

The debates over news reporting, dietary recommendations, and advertising claims illustrate the dilemmas involved in communicating risks to the public. Though the context of communication differed, the debates all centered on questions about the strength of technical evidence, the reliability of public statements, the timing of disclosure, and the dangers of causing undue anxiety. But often in these debates technical issues about the extent of risk served as surrogates for even more contentions and questions of political and social control.

Such issues of control emerged in the coffee controversy where debate involved not only the level of certainty about the coffee-cancer link, but also the questions of the proper relationship among researchers, the scientific peer review system, and journalists. How should control over risk communication be distributed among these groups? Who should control decisions about what is published and communicated to the public? What is the public's right to know?

The dispute over dietary recommendations involved issues of control within science. When experts are themselves divided about the meaning of evidence and its adequacy for policy, who has the authority to speak for science? Which interpretation should be granted the credibility of NAS backing? Whose interpretation of evidence should be communicated as the "truth," and who will be marginalized as holding a minority position? In a rapidly moving and controversial field, after all, the consensus is likely to change. In fact, by the end of the 1980s there were three new reports by government agencies, all recommending reduced fat consumption, and they elicited little dissent.

The controversy over health claims triggered by the All-Bran ads centered on questions of regulatory control over industry decisions. What types of claims will companies be permitted to make about their products? How will they be regulated? How much evidence is required to substantiate a health claim?

Practical rules or general formulas for resolving such questions are unlikely to emerge. Most everyone may agree with broad statements: that risk communicators should "avoid undue alarm," that they should "not withhold necessary information." But when such admonitions are applied to specific cases, conflict is inevitable. For communication disputes reflect the fundamental social tensions that divide consumers from producers, scientists from journalists, and groups with conflicting ideologies and economic stakes.

Let me conclude with a brief postscript on certain paradoxes that I see in the concerns about risk communication. They are, as noted above, predicated on assumptions that communication will profoundly affect consumer behavior. But what, in fact, is the affect? Studies of media influence are generally inconclusive. In particular, consumer responses to information about dietary risks suggest some bizarre ironies. On the one hand, both advertisers and the media have responded to consumer interest; dietary information has proliferated. And clearly, industry is sensitive to the influence of this information. But what in fact has been the response? There is a growing scientific consensus about the benefits of reducing fat consumption; the dietary guidelines in the late 1980s elicited little dissent and were very widely disseminated. But in a paper forthcoming in the *Journal of the National Cancer Institute*, Marion Nestle pulls together data suggesting their limited affect on marketing and consumer habits. The national food consumption surveys, coming from different agencies, disagree about the changes in calorie consumption from fats. But, according to the Economic Research Service, the availability of fat in foods has risen steadily.

While consumers have substituted skim for whole milk, there is increasing consumption of cheese, packaged foods, and other hidden sources of fat. Advertising of fast food and quick snacks, heavy in fat content, has expanded, for people no longer eat—they graze. Recent sales for potato chips and related snacks were up 23%. In 1989, according to the Food Marketing Institute, 9000 new food products were introduced into the market, including 1300 candies and snacks, 1300 dairy items, 1100 bakery products, 500 processed meats, and 118 new breakfast cereals. Many, of course, may replace existing products and many will fail to catch on. But the proliferation of such products raises a number of questions.

Is the deluge of conflicting information simply ignored by the public? Or are people blissfully unaware of risk debates, relying on the quick and easy information presented in ads to shape their diets? Perhaps people are fed up with blaming themselves for disease, or discouraged with the difficulties of making rational choices when so many sources of fat are difficult to discern. Thus, while fear of risk may be a form of "cancerphobia," the widely expressed concern about risk communication may reflect a kind of "mediaphobia"—an irrational fear that the public will respond to information in predictable and problematic ways.

This paper was developed from research in collaboration with Stephen J. Hilgartner of Columbia University.

CHAPTER 14

Nutrition and Cancer: An Historical Scientific Perspective

John Higginson

Few areas in human health generate more controversy than dietary recommendations in relation to such chronic diseases as cancer. The constituents of the "optimum diet" have remained elusive for more than 50 years. It is assumed that our scientific knowledge on these issues has followed a logical and rational progression; however, the available data have continued to be inconsistent and open to alternate explanation. This is equally true whether considering dietary factors as either promoting or inhibiting carcinogenesis. Thus, the opinions on the ill effects or benefits of certain dietary components tend to be strongly held, because they are influenced by personal conviction and other considerations rather than critical analysis.

The major dietary issues relating to cancer over the last 60 years include:

- the probable role and potential impact of ingested carcinogens
- the general impact of different dietary patterns, including both macro- and micronutrients, on cancer incidence in various cultures and environments
- possible control of cancer through dietary manipulation

Two major lines of research developed: (a) studies on the identification of carcinogens (especially man-made) in the diet, and (b) evaluation of the impact of nutritional deficiencies on cancer.

My own views are based on my experience in Africa, Asia, and the United States and modified later by my perspective as director of an the International Agency for Research on Cancer where I developed a multicountry program on these problems. Certain individual cancers will be discussed for illustrative purposes, but these remarks could be equally applicable to other cancer sites. My objective is to illustrate how uncertainties in interpreting the scientific data have caused difficulty in providing the public with definite dietary recommendations.

THE ROLE OF INGESTED CARCINOGENS

The list of chemicals found in food that are carcinogenic to animals has grown considerably since the early thirties. These include both xenobiotic and naturally occurring chemicals, usually in very small quantities. Considerable attention was also paid to carcinogens in smoked fish, superheated fats, burnt meat, and so on, without establishing convincing causal associations with cooking methods.

In the 1970s, a new player came on the scene—the *N*-nitroso compounds. These are a range of powerful carcinogens affecting many sites in animals. They can also be formed internally in the stomach or mouth from secondary amines, nitrates, and nitrites, all of which occur in normal diets. Further, they may occur at increased levels in some types of preserved foods such as meat or fish. Despite considerable research, their role in humans remains unconvincing, except possibly for the stomach.

In addition to the *N*-nitroso compounds, Ames and others have drawn attention to the large number of natural carcinogens (as well as inhibitors) that occur in food and pose theoretically greater carcinogenic hazards (as well as benefits) than man-made chemicals. Since no one knows how to deal with these, they are virtually ignored by the FDA and other authorities. However, numerous studies in humans have failed to show that such ingested animal carcinogens have had any significant impact in humans despite their theoretical potential, with the exception of aflatoxin and liver cancer in Africa and Asia and ochratoxin in the Balkans. These are both naturally occurring fungal contaminants.

Recognition of carcinogenic chemicals in diet motivated much of the deliberations of the joint WHO/FAO committees on food additives in the post World War II period. Thus, regulatory effort was and has been directed to the prudent control of unnecessary exposure to such potential hazards. Concern in the United States culminated in the 1958 passage of P.L. 85-929 on the addition of animal carcinogens to food. This law, widely known as the Delaney Clause, essentially requires that no substances carcinogenic to animals can be added to food. This implies zero tolerance, which with improved analytical methods, has caused problems. The law exempts naturally occurring carcinogens in food.

However, the view that direct carcinogens in food are important in human beings has become increasingly doubtful as the number of possibilities of exposures was found to be very great and sometimes unavoidable. For example, pyrolysis of amino acids makes them mutagenic and often carcinogenic to animals. In 1982, the National Research Council concluded that there was no conclusive evidence that such contaminants were important factors in human cancer.

Nonetheless, the perception among the general public of the role of such contaminants has been very different, as illustrated by Proposition 65 in California and the infamous Alar scare. Alar, a chemical with valuable properties, has been used for 25 years without evidence of harm. Much of the United States public including politicians, however, accepted a report, based on biomathematical models and assumptions, that was launched through a sophisticated media campaign by the National Resource Defense Fund (NRDF). The major claim was that children were at high risk of developing cancer. The public reaction was violent and hysterical and could not easily be assuaged by public officials. In 1985 the Scientific Advisory Panel (SAP) of EPA had advised that although Alar posed no imminent hazard, further research was needed. Additional studies were negative. Less than four angiosarcomas of the liver, the most probable "sentinel" cancer, are registered each year in the United States SEER (Surveillance Epidemiology End Results) population sample covering more than 25 million persons. These rates have remained essentially unchanged since the early 1960s.

However, an NRDF scientist in a teaching workshop expressed the view that NRDF's role was to raise concern on the issues, and not to present a balanced picture. While the removal of Alar from the marketplace was no great catastrophe except to apple growers and those who like unblemished and cheaper fruit, there is concern that it may represent the beginning of simplistic attacks on many other commonly used useful chemicals without a critical and balanced analysis of the issues.

NUTRITIONAL DEFICIENCIES, CARCINOGENS, AND CANCER

With recognition of widespread nutritional disease in the post World War I period, there was great emphasis on provision of an adequate diet with resultant recommendations of the amount of nutrients needed for good health. Such standards were established by the League of Nations. Nutritional scientists in the 1930s tended to concentrate on the treatment of certain specific syndromes, for example, scurvy, beriberi, and pellagra, that were caused by deficiency of a specific nutrient, the replacement of which was rapidly effective. The attack on nutritional deficiencies was basically hard science with clear-cut research results and effective prevention. The complexities of the role of diet in cancer remained poorly understood, as illustrated by the history of the causes of human cancer.

In the early 1920s, a dye, "butter yellow," was found to produce liver tumors in rats that were riboflavin deficient. However, this deficiency was carcinogen specific and irrelevant to the human situation. Nonetheless, for many years, vitamin B deficiencies were considered important in human liver cancer.

Around 1950, studies in England and Canada suggested that choline deficiency led to cirrhosis and cancer of the liver. In Canada, choline actually was added to alcoholic beverages as a preventative measure, although choline deficiency is almost impossible to produce in humans. The deficiency hypothesis was strengthened by the rediscovery of kwashiorkor, a disease of children in West Africa, by Western scientists. Such children suffered severe protein caloric deficiency and had fatty livers regarded as equivalent to the choline deficiency livers in rats. Accordingly, it was accepted that this disease was the basic cause of African liver cancer. For nutritionalists conditioned by successes in administering vitamins for beri-beri, scurvy, and pellagra, the concept of a single deficiency (i.e., protein deficiency) as a cause of cancer was attractive.

Distinguished scientists in North America and Europe, including the American Cancer Society and the National Cancer Institute, perpetuated the idea, although growing data from Africa indicated the suggested sequence did not occur. The neatness of the hypothesis based on animals outweighed the contradictory data in humans. Studies in the 1940s and 1950s indicating that cancer was inhibited by starvation also were ignored.

Over the last two decades, the situation has become clearer and today it is widely accepted that the major cause of liver cancer in North America and Europe is excessive ethanol ingestion, although it is not an animal carcinogen. In Africa and Asia liver cancer is believed to result from a mycotoxin called aflatoxin which contaminates stored grain and peanuts. However, a synergistic action of hepatitis B virus appears necessary. The saga is not yet over. A report this year of a large study in China sponsored by the National Cancer Institute (NCI) concluded there is no correlation between aflatoxin intake and liver cancer, and proposed a new hypothesis based on "dense" nutritional intake. In contrast, within a month, another report from China, also supported by the NCI, and based on molecular biological techniques, produced powerful evidence implicating aflatoxin. Aflatoxin, by the way, is common in United States grain but is not considered a significant hazard in practice.

OVERALL NUTRITIONAL PATTERNS

Originally it was believed that a carcinogen could induce cancer through a single critical event. Since 1940, however, there has been growing evidence that for most common cancers in humans, carcinogenesis is a multistage process.

In 1950 the view was expressed at a famous meeting in Oxford that diet and nutritional factors were important modulators of many types of cancer. This meeting instituted the modern era in environmental carcinogens. Kennaway said that changing any component in diet, such as an amino acid, could modify the carcinogenic process. Based on our own studies in South Africa, we concluded that the majority of cancers (about 70–80%) had an environmental factor. This concept included certain occupations, tobacco use, and other lifestyle factors of which diet was considered the most important. At that time, we were in no position to make definitive recommendations as to specific nutrients. However, our views were misinterpreted as applying only to man-made chemicals, although the original report had clearly emphasized dietary patterns and macronutrients.

Later, we and others investigated cancers of the stomach, large bowel, breast, endometrium, ovary, and so on. Most studies were inconclusive, but some suggested meat and fat might influence digestive tract and breast cancers. There were, however, many inconsistencies. For example, meat intake did not explain the low incidence of cancer in Mormons as compared to the general United States population.

Today it is believed that the balance of a wide range of normal dietary components such as amino acids, saturated or unsaturated fats, and sugars, acting as promoters, enhancers, or inhibitors are involved in cancer induction. However, despite numerous epidemiological studies, relatively little new had appeared by 1980, and the flimsiness of many early hypotheses was apparent. Although numerous efforts have been made to extrapolate more accurately to humans, animal studies have, in fact, added little understanding in terms of either macro- or micronutrients in the diet because experimental diets tend to be based on administering nutrients at almost pharmacologic levels. Accordingly, the National Research Council organized two committees to report on the role of diet and develop a consensus. The first report appeared in 1980 and covered all chronic diseases, and the second, in 1982, only discussed cancer. The first, entitled *Toward Healthful Diets,* concluded that the data available were largely insufficient to make concrete recommendations.

The 1982 report was written by a different, but also distinguished, group of scientists and provided the most complete compilation of the literature available on diet and cancer to that time. They stated that little evidence was found that contaminants or additives to the diet had any impact on human cancer and concluded that:

> The evidence reviewed by the committee suggests that cancers of most major sites are influenced by dietary patterns. However, the committee concluded that the data are not sufficient to quantitate the contribution of diet to the overall cancer risk or to determine the percent reduction in risk that might be achieved by dietary modifications. (Ch. 18, pp. 18–19)

The conclusions given above in fact did not significantly differ from those expressed by the original committee. However, The National Academy, in a summary to the public, made a number of definitive suggestions as to the benefits to be anticipated from specific changes in American eating habits.

In my opinion, the data published between the 1980 and the 1982 NAS reports were not sufficient to justify significant differences in scientific evaluations or the definite recommendations made by the National Academy. Rather, I believe the 1980 committee may have been unduly conservative toward the role of diet, whereas the 1982 committee reflected a more optimistic viewpoint. However, the differences in style and emphasis led the public to believe new data were available providing stronger proof of association.

It is worth examining the background of these variations in opinion in more detail. Considerable funds were and are being spent studying the role of diet and cancer. It is extraordinarily difficult to admit to spending large sums of public money without any firm conclusions. Another factor was the success of certain interested and influential bodies, for example, the National Institutes of Health and the American Cancer Society, in emphasizing that health can come through better dietary practices. While the good intentions of such recommendations is obvious, their scientific backing relative to specific dietary practices and cancer has been and remains somewhat unconvincing. Today, the results of many earlier studies, although accepted at the time, have been increasingly criticized on methodological grounds. More recent attempts to quantitate benefits of dietary change remain uncertain and partly speculative.

Another example relates to intake of specific macro- and micronutrients, which illustrates that even if there is a general consensus, there may be differences in interpretation and controversy among scientists. Naturally, this translates into difficulty and confusion in communicating recommendations to the public. It might be thought that in developing dietary guidelines, the amount of vitamins, proteins, and so on, necessary for good health would not be controversial. Over the years, RDAs for certain vitamins were published every five years by the NRC. The committee established in the mid-1980s, however, was given the charge of starting with a clean slate and taking a new look at previous conclusions. It thus proceeded to develop new guidelines based on a modern evaluation of the original data. In two cases, this led to lower RDAs than those previously recommended. You might think the committee would be commended for its zeal in modernizing and updating data base. Not at all! Certain influential groups, notably the Food Research and Action Center (FRAC), without seeing the report, issued the following statement in a press release (September 23, 1985):

> These consequences would be particularly shocking at a time when reports continue to show increases in poverty, especially among children, and when evidence of increasing hunger is widespread. We fear that decreased RDA's will be used to "prove" that fewer people are hungry in the United States. It would be very convenient at this time to wipe out hunger with a simple change in the numbers.

FRAC predicts that if the proposed changes are made, the new standards will be used as a rationale for decreasing funding for local school lunch and breakfast programs, food programs for day care centers, the summer food program, the elderly nutrition program, and the food stamp program.

It is clear from the discussion that developed that a major part of the argument was related not only to school meals, and so on, but possible cancer inhibition. The journal *Science* got into the act and stated that it was undesirable for the public to receive conflicting signals in areas of scientific uncertainty. The journalist did not bother with scientific validity, but concentrated on the political story. He wrote that the distinguished chairman of the scientific committee was a member of the "old guard" because he recommended rigorous and objective scientific evaluation. He implied that the 1980 NRC dietary report was influenced by industry and that scientists should be more concerned with the social implications of their work than with the facts. The correspondence among the Academy, the committee, and others is depressing and indicates how far nutritional recommendations had moved toward satisfying public and political requirements rather than reflecting scientific methodology. This misinformation, perpetuated by such an influential magazine, is not a very encouraging basis for developing effective public health recommendations for the future.

THE INTERNATIONAL AGENCY FOR RESEARCH ON CANCER (IARC)

The IARC presents what probably is the most objective evaluation on potential human carcinogens through its monograph program. Over 20 years ago, the Agency began to evaluate chemicals potentially carcinogenic to humans through use of expert groups to review the strength of the human and animal evidence. This has proved to be an effective program and the IARC monographs are widely used by more than 50 national governments, including the United States, as well as by individual public health authorities and scientists. It is assumed that because Lyon is many miles from Washington, the results are objective and unbiased. In practice, international committees have their biases and are equally susceptible to the views of a strong chairman or dominant members, as illustrated by reports on a recent IARC workshop on certain beverages. When considering coffee, the general view before the meeting was that there was no convincing epidemiological data that it is a cancer risk to humans. However, at the meeting there was a strong push by some individuals to describe coffee as a human carcinogen. This was strongly opposed by others. Because the group could not come to a consensus, no evaluation was made. Nonetheless, reports from the meeting indicate how scientific background influenced the arguments and opinions that pitted the experimental scientist against the epidemiologist.

CANCER PREVENTION

In the 1940s experimental evidence was produced suggesting that undernutrition, through inadequate intake of calories, protein, fat, and so on, could inhibit cancer induction and increase longevity in animals. For a time, this theory lay dormant. Then began a swing in views on cancer causation to a belief that some undernutrition, especially in childhood, may be beneficial. With cancer as with heart disease, fat bouncing babies were out. It was postulated that widespread protein and calorie deficiency in infancy might inhibit breast and endometrial cancer in Africa. At the same time there was evidence that the global decline in gastric cancer was related to increased use of dairy products, which, however, did not affect cancer at other sites.

In recent years there has been growing enthusiasm for preventing cancer through a variety of cancer inhibitor agents in the natural diet. These include vitamins A, C, and E, carotene, green and yellow vegetables, dairy proteins, selenium, and fiber. Although the theoretical basis is sound, the human data for most factors are far from convincing. It is almost impossible to construct a diet high in vitamin A without also being high in green vegetables. The latter contains a number of known cancer inhibitors, which may confound interpretation.

In rodents, normal components of diet, such as fructose and sucrose, have been identified as promoting cancer, and presumably, under certain circumstances, could have similar effects in humans at excessive levels. Many believe it is not fat or meat that affects cancer, but rather total calories. Unsaturated fatty acids are believed to be good for the heart, but there is experimental evidence that they are more carcinogenic than saturated fats. Certainly recommended changes in diet of this nature such as increasing vitamin A or decreasing fats may do no harm, but there is no proof that such changes are beneficial. In fact, possible harmful effects of excessive vitamin A cannot sometimes be excluded. Nonetheless, since the formation of *N*-nitroso compounds from secondary amines can be inhibited by vitamin C and other antioxidants, there is some logic to their use as cancer inhibitors where such a cause is suspected.

In conclusion, over the last 20 years there has been increasing interest in exploring agents that prevent cancer prior to its onset, or that inhibit its development and progression. Many nutritionalists regard such an approach as being justified by a number of new studies. However, to date much of the data remain unconvincing. My own views are well expressed in a recent editorial in the *Lancet* (1991), entitled "A Carrot a Day Keeps Cancer at Bay?"

> "Whilst the broad hypotheses of epidemiology lend further credence to the popular notion that there must be something in nutrition, epidemiological methods might be too crude to come close to identification of the causal substance. Only a handful of mostly synthetic substances, whose mode of action was at least partly understood from the laboratory, have shown proven efficacy for specific dermatological indications, and these often in doses close to the toxic; so many a large trial on the effects of another lower dosed and naturally occurring member of the vitamin A family to prevent other types of malignant disease, as seemed warranted by observational research, may have been started in vain. Once, during a heated debate on funding priorities of a national cancer society, a toxicologist was overheard to exclaim passionately that ". . . nutrition *must* be important, there are tens of thousands of chemical substances in our food . . ." Whereupon a molecular biologist dryly quipped: "Yes, tens of thousands, that is why we do not study it . . ." The difference between research generated by specific hypotheses (Medawar's "art of the soluble") and the more or less hypothesis-generating studies of nutritional epidemiologist, who are at best looking for very weak associations, could not have been stated more succinctly. The outcome of the debate will determine whether we will stick to the cautious general admonition to vary our food intake so that we regularly eat vegetables of all colours, or whether the 21st century will see us all swallowing preventative pills containing the winning competitor from industry."

Willingness in the past to accept a logical theory contrary to the human data might not be surprising considering the mindset of the period, but is not acceptable today as we are faced with enthusiastic but conflicting claims as to the cancer-inhibiting properties of individual food components.

The question remains, at what point do scientist become advocates? Could the drive to increase research funding unconsciously urge some scientists toward a more optimistic interpretation of the data?

I wish to emphasize that while accepting an important role of nutrition in cancer, the jury is still out regarding the role of many individual components. All too often the scientific opinions expressed by articulate individuals represent strongly held views that are sometimes ideological and political, occasionally with some unconscious self-interest. Although there appears to be little evidence that the multitude of potential animal carcinogens and mutagens present in the average diet represent a significant risk to humans, the public perceptions of such risks strongly influence national and international regulations. Final decisions and recommendations will depend on sound scientific judgment and a weight of the evidence approach, and this will apply equally to biotechnology. For the moment, recent research has added relatively little to former recommendations on dietary balance and avoiding overeating.

SOURCES

Ames, B.N. (1983). Dietary carcinogens and anticarcinogens, oxygen radicals and degenerative diseases. *Science*, 221, 1256–1264.

Clemmesen, J. (Ed.) (1950). Symposium on Geographical Pathology and Demography of Cancer, session held at Regent's Park College, Oxford, England, July 29 to August 5, 1950. Copenhagen: Council for the Coordination of International Congresses of Medical Science.

Delaney Amendment (1958). Food Additives Amendment of 1958 to the Federal Food, Drug, and Cosmetic Act, Public Law 85-929, Sept. 6, 1958.

Doll, R. and Peto, R. (1981). The causes of cancer: Quantitative estimates of avoidable risks of cancer in the United States today. *J. Natl. Cancer Inst.*, 66, 1192–1308.

Editorial (1988). A carrot a day keeps cancer at bay? *Lancet,* 337, 81–82.

FRAC (Food Research and Action Center) (1985). Proposed changes in National Academy of Sciences nutrition standards could mean more hungry Americans (news release), Sept. 23, 1985.

Geboers., J., Joossens, J.V. and Carroll, K.K. (1985). Introductory remarks to the consensus statement on provisional dietary guidelines. In *Diet and Human Carcinogenesis*, J.V. Joossens, M.J. Hill and J. Geboers (Eds.), New York: Excerpta Medica, pp. 337–342.

Graham, S. (1983). Toward a dietary prevention of cancer. *Epid. Rev.*, 5, 38–50.

Higginson, J. and Sheridan, M.J. (1991). Nutrition and Human Cancer. In *Cancer and Nutrition*, R.B. Alfin-Slater and D. Kritchevsky (Eds.),. New York: Plenum Press, pp. 1–50.

IARC (1991). IARC Monographs on the Evaluation of Carcinogenic Risks to Humans, Coffee, Tea, Maté, Methylxanthines (caffeine, theophylline, thiobromine) and Methylglyoxal, Volume 51. Lyon: International Agency for Research on Cancer.

NAS (Committee on Diet, Nutrition, and Cancer, National Academy of Sciences) (1982). Diet, Nutrition and Cancer. Washington: National Academy Press.

O'Neill, I.K., Von Borstel, R.C., Miller, C.T., Long, J. and Bartsch, H. (Eds.) (1984). *N*-Nitroso Compounds: Occurrence, Biological Effects and Relevance to Human Cancer, Proceedings of the VIIIth International Symposium on *N*-Nitroso Compounds held in Banff, Canada, 5-9 September 1983, IARC Scientific Publications No. 57. Lyon: International Agency for Research on Cancer.

Surgeon General (1988). The Surgeon General's Report on Nutrition and Health, U.S. Department of Health and Human Services, Public Health Service, DHHS (PHS) Publication No. 88-50210. Washington: Government Printing Office.

Tannenbaum, A. (1945). The dependence of tumor formation on the degree of caloric restriction. *Cancer Res.*, 5, 609–615.

WHO (World Health Organization) (1958). Procedures for the testing of intentional food additives to establish their safety for use. Second report of the joint FAO/WHO Expert Committee on Food Additives. *W.H.O. Tech. Rep. Ser.*, 144, 1–19.

Willett, W.C. and MacMahon, B. (1984a). Diet and cancer. *N. Engl. J. Med.*, 310, 633–638.

Willett, W.C. and MacMahon, B. (1984b). Diet and cancer. *N. Engl. J. Med.*, 310, 697–703.

Williams, G.M. (1988). Sweeteners: Health Effects. Proceedings of an International Conference Sponsored by the Environmental Health and Safety Council, American Health Foundation, February 18–20, 1987, New York, NY. Princeton, NJ: Princeton Scientific Publishing Co.

CHAPTER 15

Diet Risk Communication: A Consumer Advocate Perspective

Ellen Haas

There is a political question in risk communication because we all come at the issues of the risks in food from very different perspectives. As one who pays the money, but does not want to take the chances, I think that my perspective as a consumer advocate is going to become quite clear.

It also is clear that the issues that Dorothy Nelkin raised in her paper are important ones for us all to consider because they are the ones that are politically charged. If we were just to have scientific findings in the laboratory and nothing was done with them—in other words there were no announcements about those findings, there were no hearings on those findings in Congress, there were no "60 Minutes" to do a show on them—we would not be here talking about the politics of risk communication.

But the real essence of the issue for the food industry, and it is also the essence of the issue for us in the consumer community, is what do we do about the knowledge that we have and what do we do about the uncertainty that does exist about the relationship of diet and health and the relationship in particular of diet and cancer?

When I read Dorothy Nelkin's paper, I realized that communicating risk is itself a risky business. But it is also an essential vehicle for anyone involved in the making of public policy. As consumer advocates, we are conduits of information to the public, watchdogs of federal agencies and the industry, and a voice for consumers on issues that have significant impact on the quality of their life. It is risky business out there because often there is insufficient substantiation of risks, and also there is irresponsible portrayal of those risks in the media that can cause unnecessary alarm or, as often happens, a whitewashing of what could be a major public health problem.

Communication is at the root of Public Voice for Food and Health Policy's mission, communicating for and with the consumer on critical food safety and nutrition issues.

While communication is the tool we use to get information to consumers, information is also a tool consumers use to make critical decisions about diet and health when they go shopping in the supermarket or when they make decisions on what to feed their family.

Risk communication is also a highly effective and provocative way to foster changes in policies that are obsolete, or to implement new policies when current ones are wanting. It is used by the food industry, consumers, and public health organizations to encourage changes at the federal level. Federal policies governing food safety, nutrition, and health should be designed to provide the best possible protection for consumers, taking into account all possible public health risks.

At the center of the debate surrounding risk communication is the formation of policy about cancer risks. In particular, the issue is how rational policy involving cancer risks can be developed when scientific information on those risks is incomplete and uncertain.

Everybody has their own version of "science." As Professor Nelkin pointed out in her paper, science is subject to change. Risk communication and the resulting public policy can then not be based on sound science alone. The reality is that we often have to make decisions about risk communication in an environment of uncertainty, and interpretations of risk are influenced by economic, social, ideological, ethical, and policy considerations as much as the scientific documentation.

Professor Nelkin raises the issue of uncertainty that automatically surrounds risk concerning cancer. But in several places she draws different conclusions than the consumer community does about what should be done in the face of such uncertainty. In her examples about diet and cancer, particularly the coffee case, Professor Nelkin suggests that the politics of risk communication is driven by the preliminary incomplete character of the research evidence, and that future data-gathering efforts would remove this uncertainty that plagues risk communicators and policymakers.

Communication about risk of cancer as well as policy responding to that risk is almost always going to be characterized by uncertainty. Communicating cancer risk is an inherently political and value-based decision for the very reason that we are unlikely ever to know with certainty the extent of the cancer threat to human beings. The current procedures for testing for cancer effects, and for assessment of human health risks of animal carcinogens, are such that most evidence that a substance or food causes cancer is by definition going to remain preliminary. We have seen politics at work as industry and consumer groups handle uncertain findings in different but very deliberate, organized, and highly polished ways. The difference can be explained largely in terms of the way the two interests respond to uncertainty surrounding the evidence.

Consumer, environmental, and public interest groups feel they have a responsibility to raise the public's awareness of potential food safety hazards so that the public can make thoughtful, informed decisions in the marketplace, and participate in an educated manner in the food policy process. Indeed, the media gave the *New England Journal of Medicine* study on coffee and pancreatic cancer widespread attention. They reported the limitations of the study and the preliminary nature of the findings, but also urged that the study be taken seriously. This is a prime example of responsible risk communication. There is no question in my mind that despite the uncertainty, risk must be made public.

Consumer groups translate the uncertainty surrounding cancer risks into public policy recommendations that would risk making an error, if any, on the side of consumer protection. This is what underlies support for the anticancer Delaney clause and phase-out of pesticides found to be animal carcinogens. Because we are in almost all cases never going to know precisely the extent of the risk that animal carcinogens pose to man, and because their elimination would not really undermine our food-producing capacity, a "better safe than sorry" approach dictates that they should be banned from the food supply.

Yet many groups that argue solely on the basis of sound scientific justification treat science with contempt when the findings of carcinogenicity tests threaten their own private interest. Their response to the uncertainty of translating laboratory results into human risk assessments has too often been driven by damage control. Thus we find, and we have seen many times in the last few years, full-page paid advertisements claiming that consumers will have to eat thousands of pounds of produce a day before they get the equivalent dosage of the carcinogen fed to laboratory rats. Or we saw that with saccharin and soft drinks all the jokes that were made about that poor rat.

Consumers are regularly bombarded with messages that try to undermine the scientific credibility of consumer and environmental groups—NRDC certainly has gotten its share in the last several years—and claims that these groups are seeking the impossible dream of a riskless society. Certainly that is industry's claim in the debate over the Delaney clause and other pesticide issues.

For all of the food industry's efforts to communicate that the American food supply is safe or the safest in the world, however, its own opinion surveys, such as FMI supermarket trend surveys, continue to show that four-fifths of the public are still seriously concerned about pesticide residues in food. We continue to see polls taken by Gallup and by Louis Harris showing the eroding confidence in the safety of the American food supply.

There is a deeply ingrained lack of trust and faith in the American people at work. Instead of leveling with consumers about the findings of animal carcinogenicity and the uncertainty of the human health implications, many in the food industry have chosen to take a reactive stance. Millions of dollars are poured into public relations efforts to persuade consumers that their concerns are not scientifically valid, that they are merely false perceptions. This failure to be honest with the public about the research findings and the limits of our understanding of cancer risk helps set the stage for the very same public eruptions that industry is trying to prevent.

Over the years, Public Voice has had particular experience in documenting the risks associated with eating contaminated uninspected seafood. There is hard evidence that the FDA is too lax in its inspection efforts; that while the agency has only set one tolerance for chemical contaminants, hundreds of contaminants are in the waters; that consumers are becoming sick and often seriously ill and some are even dying from eating contaminated uninspected seafood; and that many of the risks are long-term, including cancer and second generation developmental problems.

In January of this year, the National Academy of Sciences released a two-year study called *Seafood Safety*, essentially concluding that the risks associated with eating contaminated uninspected seafood are multifaceted, multiregional, and affect all types of fish, and that current federal and state inspection programs are inadequate.

Yet, there are still some who use the study to support their claims that most of the problems are just acute; there are no cancer problems, or they are just associated with raw shellfish, and, therefore, we should target areas where we know of an absolute risk rather than waste government dollars. Since the extent of risks associated with eating contaminated uninspected seafood are uncertain (we know, however, that people are getting seriously ill, according to the Centers for Disease Control), consumers do need to be alerted and public policy needs to be reformed to take in mind the necessary public health protections.

It is not the purpose of risk communication to dictate to consumers what to eat and what not to eat. Risk communication really is a kind of freedom of information for the public. It is the right to know about potential health risks. Industry, scientists, and to some degree the Federal Government too often are looking for unrealistically strict validation of data before any public communication is warranted. If such criteria becomes the *modus operandi* of assessing risk communication, consumers will be asked essentially to bear risks of which they are completely unaware. In the end, the public will be rendered incapable of making thoughtful decisions about the acceptability of the risk.

It sounds like an Orwellian fairy tale, where Big Brother gets the first and last word on what the public will or will not know. I opt for a more enlightened independent populace, rather than George Orwell's divine ignorance of the day.

We live in a democratic society where the public is eager to participate in policy decisions, and the government is willing to encourage and does encourage such participation. Our democracy also prides itself on the free constant flow of information to its citizens, through which they can become enlightened. In a world of ivory towers, sound science may rule the day. But in today's real world of uncertainties, politics, economics, and, most importantly, personal values also play a part in policy decisions related to food safety.

Since communication concerning cancer risk is kept constant by uncertainty, let us risk telling the public all there is to tell. What consumers do not know about cancer risk may hurt them, but what they do know can only help us deal effectively and pragmatically with those risks.

DISCUSSION OF PART IV:

MR. HIRSCH: I had the feeling, listening earlier today, that the promised great biotechnical revolution in agriculture will not answer our greatest and most immediate needs. The problems of providing food are largely social and political.

I have also had the feeling that we are misled into thinking that the great issues linking the food supply and health have to do with pesticide residues. I think Professor Higginson put his finger on something vastly more important. We heard last night that perhaps one in a million people may be affected by the adverse affects of pesticides, but 70,000 others have cancer. I am very concerned about the role of the food supply in determining what happens to the other 930,000 of the million. All will die, and the question is, how they are going to die; with what degree of osteoporosis, with how much Type II diabetes, with how much hypertension? There is much evidence linking lifelong food intake patterns with many of these diseases. Regrettably, a lot of the evidence is reminiscent of that study in the *New England Journal of Medicine* linking coffee to pancreatic malignancy. It is correlational and epidemiological. It is the weakest link in the lines of scientific evidence that lead to final conclusions on which actions should be taken.

I believe we are now at the dawn of a biotechnical revolution. What we are going to see in the next few decades from the agriculture and the food industry is the opportunity to make foods the likes of which we have not dreamt of in the past. If 30% fat is healthful, then we can make low-fat foods that are tasteful and abundant. This can be done inexpensively and distributed equitably. Food availability will not be dependent on economic motives alone. We will now have the wherewithal to tailor-make the foods we need and make them available in abundance.

We are at the dawn of a quantitative biologic science, which we have not had before. The coming of the revolution in molecular biology is going to permit us to look at nutritional diseases in a way that has not been possible in the past. It will be possible to look at differences in human genotypes and thereby differences in individual susceptibility to different foods. We will be able to examine early developmental stages by studying the effects of food intake in infancy and then the later effects of the early years to the later emergence of food habits and illnesses.

My wish is that a partnership could be arranged among all of you here, representing industry and government, and this new nutritional research. For many years, nutritional research has been descriptive. But now we can solve problems, and, with industrial support to sustain the nutritional sciences, begin to get the answers the public needs.

Lastly, to Ms. Haas, I think the public should be fully informed, but we regrettably do not always have enough good information to give them. Now the information will be forthcoming. As of this moment, all dietary recommendations should be in the category of "prudent diets" or current and tentative recommendations. We do not yet have a firm scientific basis for most dietary recommendations, but we will.

PROFESSOR NELKIN: There is a political view suggesting that the financial commitment to research on individual predisposition to risk is a way to divert attention away from environmental carcinogens, by placing blame on the individual. That kind of a view indicates the basic mistrust that dominates, that pervades, the whole discussion of risk.

MR. HIRSCH: A paranoia that perhaps we deserve, but it is paranoia nonetheless.

MS. HAAS: In a discussion about diet and cancer, you said, it is more important to look at the problems of osteoporosis; and Dr. Higginson said let's look at the problems of poverty, those are the most important. I hear that very often, when we are talking about cancer risks, what we should really be looking at are the microbiological problems.

We are not talking really about risk trade-offs. When we are dealing with diet and cancer problems, I think we have to deal with those issues and not digress. When I first started my consumer advocacy about 18 years ago, I would go and testify on auto repair because that was a major problem. Invariably, the auto repair lobbyists would come up to me, "You know, you should really get after the TV repairmen; they are the worst, or you should get after the insurance industry. Anybody but me."

In the area of cancer, the American consumer has a particular concern. It is not a phobia, it is not just a fear; it is a concern when 400,000 Americans die each year, and the public wants to have as much control as possible. Again, you are trading off economic interests against health interests. I have to take exception to talking about other issues rather than dealing with the one that we have.

MR. HIRSCH: I absolutely agree with you, and I did not mean to trade off. I think it is a fundamental issue of nutrition and its relationship to cancer that must be looked into by investigative techniques that are now becoming available.

PROFESSOR HIGGINSON: I would like to emphasize that I believe it is a question of priorities. As Bentham said, because a major risk exists, there is no reason to ignore a minor risk. On the other hand, if you are looking at the greatest good for the greatest number and you have limited resources, I think, morally, your lines of policy are fairly clear cut, of course. You may disagree what is minor. We are really talking most often of informing the public of a potential of a very minor risk that is distant and undetectable. Yet it is extraordinarily difficult to tell people that they could ignore such risks for practical purposes and live happier.

I believe that further research is necessary, for example on osteoporosis, which illustrates the problems of "good" science. I have just come from Africa, where osteoporosis in the black population is a tenth of that in U.S. whites, and their calcium intake is far below that of the white communities here and abroad. Yet, here we are feeding calcium to prevent this disease with little effect. I am not saying that osteoporosis is unimportant, but that we do not know the answer.

We still do not know what is an optimum diet. If we really wanted to start again and have effective prevention, we could all become Seventh Day Adventists. We are starting late in life and the benefits will not be immediately appreciable, but our children will benefit because Adventists' cancer rates are around 60% of that of the U.S. population as a whole.

MS. CULLITON: We have been talking mostly about biotech kinds of risks all day. We have been talking about the things in the food supply that might or might not cause some harm to somebody somewhere along the line.

Then you brought up a different kind of risk altogether; namely, the contaminated seafood. Now, if I understand what you are saying correctly, if we were served contaminated seafood this evening, most of us would be sick by morning. That is a different kind of risk from the question of whether or not I should have an additional cup of coffee.

You must receive these kinds of questions from unscientifically informed citizens all the time. How would you explain the difference in risk between eating contaminated seafood that is going to make everybody sick and changing your habits with respect to drinking coffee?

MS. HAAS: On the problem of contaminated seafood: you might get sick in the morning, but you might also not know the consequences for 20 years, because as much of the problem could be the chemical contaminants and the fact that there are not safe levels or tolerances established. Getting sick in the morning is only one example of the result of contaminated seafood. You might have other results from the contamination.

I think our basis is two things. Our focus, at least at Public Voice and I think at Citizens for Science in the Public Interest as well, is on public policy. It is not to tell consumers to have four cups of coffee tomorrow or to eat three portions of yogurt and stay well. Ours is to see that we have a regulatory system and public policies that assure us a safe and nutritious food supply and to fill those gaps in government protections where consumers do not have the necessary food safety.

So we would not get into those questions or recommendations specifically. That is for educators who are nutritionists to do. I have nutritionists on my staff, but we are instead really looking at the public policy holes. I think those are very different questions, because then you are dealing with the aggregate and not with just the individual.

MS. CULLITON: Would you recommend that the FDA institute some sort of regulation on coffee drinking? If you are going to make the recommendation not to the citizen but to the regulatory agency, at what point do you make that decision?

MS. HAAS: On coffee we have not taken a position. I think that what we have tried to do is to evaluate the issues to see the ones that have the most need for change and reform. Certainly the lack of a mandatory seafood inspection program is one, as is the need to reform our pesticide policies. The coffee issue is not one we have come down on.

PROFESSOR NELKIN: Barbara Culliton, you referred to seafood contamination, but I see no relationship between regulating coffee and seafood contaminants. I do not know what is behind your question. As soon as you use the word contamination, you are putting a bias on the issue. It implies either sloppy practices or deliberate neglect. There is no question in anybody's mind that food should not be contaminated. There is no uncertainty about that. In contrast, there is real scientific disagreement about the effect of coffee, and there is consumer choice. Nobody would choose to take a risk of eating contaminated seafood, where people do choose to drink coffee. You are comparing "apples and oranges," to use a food metaphor.

DR. VAN DER HEIJDEN: Dr. Higginson, I think it was in 1967 that you came out with your article stating nutrition or diet to be a very important cause of cancer. Afterwards, Doll and Peto came with the same opinion. I remember the message was that it could be chemical factors, and it was chemicals that we considered the main cause of human cancer. In 25 years there has been a tremendous scientific search and a lot of money spent to find those chemicals. We were at that time so naive that we were convinced that if we could only detect them by short-term tests or bioassays, we could subsequently eliminate them, in order to get rid of cancer.

Now, 25 years later, there is not that much progress. We did not find many chemicals that are carcinogenic to humans in our diet. You mentioned a few, but even those might not be relevant in our Western society. We considered fibers as an important factor, but there is also a lot of doubt as to whether they play a role. We considered fat; but, in fact, it is still a largely unknown area.

Maybe overnutrition is the most relevant aspect of diet and cancer. However, that does not fit into your hypothesis on poverty, because in our Western world we have the highest energy levels in our diet.

Ms. Haas, you mentioned there are polls indicating concern for food safety among large segments of the population. In Holland these polls have been done, too. The conclusion was that the majority of the population do trust government, feel safe, and have no problems. They are not specially educated, do not have much knowledge about food safety, but there is just trust in governmental institutions. A small part of the population does not care either. But they are often specialists, and they are actually educated.

But there is a small group of people with only limited knowledge who believe that food is unsafe. Those are often the trendsetters for the media. We need to provide more education and should put more emphasis on risk communication, because I feel people are still concerned about pesticides in the food, and it should not be a real hazard to fear.

MS. HAAS: I think that the concern is very real. It is on a similar track to the concern about the environment. I think what you will see in the next decade is the coming together of food safety, health, and the environment. The same concerns that have been widespread both in Europe and the United States—actually all over the world—about the environment are now coming to food as well, and pesticides and other food safety issues are very close to that.

The kind of teaching that needs to go on has to be in the context of what those risks are, and their acceptability, rather than teaching people not to worry about it, implying that the whole food supply is safe. There are hundreds of polls, and I think that there have been continual trends confirming that there is this declining confidence in the quality of the American food supply. That has come about as we have obtained more information about the risks.

The solution is not to go underground about those risks and not give the information to consumers, but rather to fill the gaps in the government protection. Again, the problem is not because some consumer advocate or environmentalist is out there saying some pesticide is carcinogenic. It is that the EPA is not doing anything about it. I think that too many have been focusing the attention on consumer or environmental groups misusing the information rather than focusing the attention on the necessary things that government has to do.

DR. VAN DER HEIJDEN: Safety issues are also here misused as a surrogate for something else, maybe quality. Food has become a fetish—the point is that the consumers do not love food anymore, as such. It is considered healthful or not, but they have forgotten that food is something to enjoy.

MS. HAAS: I disagree with you completely. I think that with the surge of interest in healthful food, you have come to see much more enjoyment. In Washington we have many new restaurants that are health and nutrition and environmentally concerned, and they are packed. When food companies started to market nutrition, there was a positive response. The more low-fat products—I think you are getting *more* enjoyment.

I disagree with you that all of a sudden the consumer does not enjoy food because now they know about the risk. Consumers who are more informed will choose the more healthful foods and then get more enjoyment out of them.

MR. FLYNN: The people of the United States are the fattest in the world and getting worse. I refuse to accept that any of this communication is having any impact. The people who want to live a more healthful life are taking care of themselves, and the other people are fantasizing.

MS. HAAS: Let me say this. I am not saying that the American public is eating the way they should. We just completed a study on school kids called "Heading for a Health Crisis." What it showed is very high fat consumption, particularly among the poorest of children. Those that participated in the school lunch program—again, you are talking public policy—had a higher level of fat consumption and poorer diets.

I agree with you that the mass population still has problems with their diet, serious problems of too much fat consumption, in particular. What I am saying is those who have chosen healthful diets are not doing it in misery, but are doing it positively. It can be fun; it can be enjoyable. I think there is a market out there for healthful eating. Otherwise, all of these food companies would not be advertising their healthful products in such a way.

MR. FLYNN: I would respect a physician who said you are going to live until 80 if your genes are not bad; now which thing do you want to die of? These discussions about solving everything are nonsensical. You know you are going to die.

MS. HAAS: It is not only a question of when you die, it is what kind of life you lead. If you lead a life with 40 years of heart disease, you are not living; you are not attaining the same kind of quality of life. The diet you have does relate to that heart condition. There is enough scientific evidence. I think that there have been enough confirming studies on the relationship of fat and heart disease to do something about our diets in the United States or in the whole world.

PROFESSOR HIGGINSON: We started off with the view that it was carcinogens in the diet that were the problem, and wasted years looking at cooking methods, and so on, for stomach cancer, colon cancer, and so forth. It was all useless. Then we studied deficiencies believing that they caused cancer. However, we have evidence going back to 1940 that ingesting fewer calories allows one to live longer, and that obesity is positively related to both cancer and heart disease.

Today we are looking at nutritional patterns with little success due to their complexity. The term poverty becomes part of the discussion because of the need for priorities in policies. A major fact in this country is that poverty is associated with an unsatisfactory lifestyle, including alcoholism, obesity, and the other features that are found in deprived populations. Even in Denmark, between certain social groups there is a difference of two to one in cancer frequencies. It is the lifestyle factors that go with poverty, not poverty per se.

DR. WOTEKI: When we were talking about the public and its risk perception, we were talking about the public as if it were a monolith, equally interested in this topic to begin with, and equally prepared to accept and understand and integrate the information that we would be providing to it.

Just a couple of weeks ago the American Dietetic Association released the results of a survey that they had done recently, a national survey, in which 40% of the respondents—they were all adults—said that, quite frankly, they were not interested at all in any of this discussion of diet and health; they could not be bothered; they did not want to hear about it. Another third of the population said, yes, they were kind of aware of it, but changing their diet would mean they would have to give up the foods that they really liked; so, therefore, they were not paying any attention to it. Then there was a quarter of the population that responded and said, yes, I follow these issues, I read the newspaper, I get information from my physician, I get information from magazines, and I use it to make changes in my diet.

So just a word of caution. I think we need to think about the public as not a monolith but, rather, as a very diverse group. We will all probably change from one group to another at different stages in life. The group that said: "I am not interested and I do not want to hear about it," tended to be younger. Perhaps when they reach middle age and their cholesterol levels go up, they will become more interested.

MR. MURPHY: I do not mind responsible reporting on information. Where I have a problem is when somebody starts moving too far into the regulatory area and determining my diet or the risks I can take and not take.

I do not know who has the right to decide whether I should be a left tackle and eat a diet that is heavy in calories, allowing me to gain the weight to play left tackle, or I should be a 100-yard sprinter, who has a different diet. Those who have played sports understand there are different dietary patterns to arrive at different body types and different work habits. We do not talk about that. It seems to me that is my choice.

I do not want somebody taking responsibility for my life. I want the information available, but I would rather exert judgment to determine the way that I want to live, including the pleasure factor of it, if that is important to me. In fact, I may want to be fat.

PROFESSOR NELKIN: Surely you would like regulation of water supplies?

MR. MURPHY: I think in the regulation area I would agree with you. I guess we are talking degree of regulation, how far you go in that area. There are some things that are obvious. I mean, where it is prompt, immediate, and disastrous. But there is a great gray area, and it seems to me I heard a lot of words—Ellen Haas in particular was talking about the right diet. There was an implication that my diet should be like her diet. I am not sure that that is appropriate for my objectives in life.

MS. HAAS: I do not think I ever said that your diet and my diet should be the same. In fact, all that we have today are dietary guidelines that are appropriate. They are not dietary edicts; they do not tell you what you have to eat. What consumers do need is information, and we did enact the Nutrition Labeling and Education Act, which will provide that information when it goes into the marketplace in a couple of years.

MR. MURPHY: I am not arguing about responsible information. I think that is fine, and I do not even mind the coffee report, because that will cause somebody else to report the other side of it anyway. I am still drinking coffee.

So it is useful. The biggest problem is when, in fact, we go so far over into the regulatory area—and I see a lot of it in our environmental movement today—that the economic tradeoffs have probably hampered other benefits we might get out of the system. We have to continually evaluate where this regulatory area crosses over both the rights of the individual to make determinations and the benefits that come from a little more freedom of action.

This country got where it has because we had a lot of entrepreneurs, working in areas that are still controversial and still open. Whether those entrepreneurs are business people or lifestyle people, they have a willingness to take whatever risk there is for a benefit they see—a benefit that you and I may not yet perceive. I am saying that freedom to make mistakes is still a very important part of our society; and every regulatory step we take bothers me, because I do feel that I would like to die the way I want to die, as best I can.

DR. ARNTZEN: As an experimental biologist, I look at regulations as guidelines that allow us to make individual decisions. One of the beauties of the technologies that we talked about this morning in biotechnology is that, in the next decade, we are going to get an incredible amount of information that will allow us to make accurate individual decisions.

The science of nutrition has been an empirical study—an observational study. But just in the last five years we have launched into an era in which nutrition is becoming an experimental science, especially in the case of animal models. It is now possible to genetically "design" model animal populations that vary in propensity to diseases so as to mimic human nutritional imbalances. In addition, molecular biology has allowed scientists to discover oncogenes and tumor suppressor genes that are important to our understanding of cancer and gene mapping. The human genome project is going to allow us to have diagnostic tools within this decade that will give us means to determine our disposition to important classes of disease and important subsets of cancer and nutrition-related diseases.

The important thing is, we are in this very emotional stage of talking about regulatory activities right now. It would be nice if we could take a forward look at how some of these new tools and technologies are going to give us the advantage of adding to our regulatory mechanisms, perhaps allowing us to come back to an individual decision-making process.

DR. SMITH: Since there is so much uncertainty in this area, the question I am asking the panel is, how will you handle a mistake? Let me give you an example. We have just gone through a torturous process in this country over the past year of claiming that tropical oils were bad for people, and we have taken them out of our foodstuffs. This has had a tremendous effect on a third-world industry. Sometime in the future we may decide that saturated fats are very good for you. Who is going to tell the populace that we made a mistake and that palm oil is good for us and we should be putting it back in our food? How would you handle a thing like that?

PROFESSOR HIGGINSON: You asked a question about a mistake; how do you deal with a mistake? In the cases of many of the risks that we are talking about concerning cancer, we would never know whether we made a mistake or not, because we should be operating at a level at which nothing would be detectable.

DR. SMITH: The example I gave was the tropical oils, because of the saturated fats and involvement in heart disease. What if we now determine that for some or most of the population, saturated fats are good? How do we go back to the population and tell them that we made a mistake?

PROFESSOR HIGGINSON: There is a whole history in nutrition of variations in various nutrients being regarded as good or bad at different times. Therefore, I have great belief in moderation. For the average guy, moderation is unlikely to result in a big mistake. On the other hand, if he embraces some fad, he certainly may make such a mistake.

PROFESSOR NELKIN: Is it problematic to tell the public that science changes, that we learn that there is no truth in science, if you like? We talk about science literacy in this country, and yet there seems to be some block against communicating the fact that science is not truth—and it is not truth—that it changes, that new information comes in. If presented in the right way, it could be looked at as an educational opportunity.

DR. SMITH: Certainly I would applaud that, but we do not tend to do that. We tend to make it absolute: "This is bad for you; we should ban it in our foods."

My point is, how do you tell the people in Thailand and those places that they are out of business now because we made a mistake—if we made a mistake?

MS. HAAS: Who you are serving is really the question. It is a hypothetical situation that is hard to imagine—what happens if we have made a mistake. Isn't it better to err on the side of public health? And if, to the best of your company's analysis, the best public health move today would be to take tropical oils out of cookies because you are serving your consumers and their public health needs, isn't that an appropriate action? You are not serving, though I have sympathy for, the developing countries whose economics are being hurt by this.

We have found out a different problem of mistakes that I think is very important and again gets into the pesticide issue. Many pesticides were approved lacking the kind of scientific data that we have today. The reregistration process has taken a very long time, we are nowhere where we should be in reregistering those old pesticides. They are the ones that are coming into the crisis situation, because the information we have today is that we based our decision-making on inadequate data. So it was a mistake in our earlier analysis that we have to go back to today.

You have to do the best you can do in your time, and not worry about what is going to happen in 50 years.

MR. TAYLOR: The one thing that is clear from this debate is that there is enormous uncertainty. Yesterday's certainties are today's uncertainties. Maybe we are going to move into an era of greater certainty and firmer conclusions. But from this present uncertainty flow some very specific consequences. If you are a villager in the Solomon Islands whose whole livelihood has been growing coconut palms for coconut palm oil, or you are a farmer in Malaysia who has built his livelihood and well-being on the back of palm oil, this is not a light issue. These are not crops that you can plow out and then plant soybeans; these are tree crops that have been there for years and years and years.I think this sort of consequence demands enormous responsibility on the part of those who make these pronouncements.

MR. HIRSCH: I have studied human obesity for the past 30 years. I can tell you just a few things relevant to our discussion. Number one, there are minimally 35 million Americans who, if they lost weight, would also reduce certain health hazards.

The medical expenditure that is attributable to obesity treatment and its hazards and illnesses is on the order of $150 billion a year. There is a $30 billion expenditure in an effort to get fat people to lose weight, which is rarely effective. There is not one shred of evidence that everything that we have done in terms of nutritional advice, new agencies, and commissions, has changed the amount of obesity in North America.

But there is very important, new scientific information indicating that obesity is a disorder of energy metabolism in which there are important genetic factors at work, certainly in the animal models, and very likely in man as well. Very important early developmental events and later psychosocial factors play a role as well. We are entering a new era in which we are recognizing that obesity is a biological disorder. Advising and telling people to do this or that, on the basis of current information, has not worked and is unlikely to work until we accumulate more information.

PART V

RISK MANAGEMENT AND RISK PERCEPTION

CHAPTER 16

Food Safety as an Element of Risk Assessment and Management

The Honorable D. Allan Bromley

One of my favorite pieces of historical writing centers on the town of Williamsburg, and now that I am here I cannot resist mentioning it. It concerns an episode in the 18th century when the commissioners of Maryland and Virginia invited the leaders of a local tribe of Indians to send some of their sons to William and Mary to obtain the benefits of a classical education in Greek, Latin, grammar, rhetoric, mathematics, and philosophy—the standard curriculum at the time.

The Indians gave the offer careful thought and, according to an 1834 record, gave the following answer:

> We know that you highly esteem the kind of learning taught in those Colleges, and that the maintenance of our young Men, while with you, would be very expensive to you. We are convinc'd, therefore, that you mean to do us Good by your proposal; and we thank you heartily. But you, who are wise, must know that different Nations have different Conceptions of things; and you will therefore not take it amiss, if our Ideas of this kind of Education happen not to be the same with yours. We have had some Experience of it. Several of our young People were formerly brought up at the Colleges of the Northern Provinces [Yale and Harvard]; they were instructed in all your Sciences; but, when they came back to us, they were bad runners, ignorant of every means of living in the Woods, unable to bear either Cold or Hunger, knew neither how to build a cabin, take a Deer, or kill an Enemy, spoke our language imperfectly,

> were therefore neither fit for Hunters, Warriors, nor Counsellors; they were totally good for nothing. We are, however, not the less oblig'd by your kind Offer, tho'we decline accepting it; and, to show our grateful Sense of it, if the Gentlemen of Virginia will send us a Dozen of their Sons, we will take Care of their Education, instruct them in all we know, and make Men of them.

One gets the decided impression that one of the products of the Colleges of the Northern Provinces must have written this reply.

The world has changed considerably since the 18th century, and in the process our perceptions of the world and of the risks presented by the world have similarly undergone a dramatic shift. A little over a hundred years ago, the leading cause of death in America's cities was tuberculosis, and when I was a child parents spent their summers dreading the scourge of polio. In 1870 a quarter of all people born had died before they reached the age of 25 and half were dead before the age of 50. Today only 3% die before age 25, and 90% live to be older than 50.

CHANGES IN THE GLOBAL FOOD SYSTEM AND THE PERCEPTION OF RISK

Equally dramatic improvements have marked the global food system. Today human beings generate more food of higher quality than they ever have in history. Furthermore, science and technology hold out great promise for future improvements in both the quantity and quality of the food we produce. In particular, biotechnology offers the possibility of progress as dramatic as that of the Green Revolution pioneered by Norman Borlaug (who is now a member of the President's Council of Advisors on Science and Technology).

A casual observer of the American scene—or the scene in many other industrialized countries—must therefore be struck by what seems to be an obvious paradox. America has the safest, best, and must abundant food supply in the world. Yet public concern about the quality of that food supply is extremely high—and seems to be going up as our food improves! Shortly after the Alar scare, a poll was taken in which 80% of respondents rated pesticide residues as a serious food hazard. Because of this public concern, over 50 food safety bills were introduced in the 101st Congress.

This concern about food safety, in part, reflects some fundamental misconceptions that people bring to their understanding of risk in our society. My own view is that many people tend to divide activities into three categories. An activity can be 100% safe—which in reality is impossible—or it can be 100% dangerous—which means it is to be avoided at all costs. Everything in between tends to be viewed as having a 50-50 chance of having unhappy consequences.

This view of probability may be an advantage to the people who run lotteries and gambling casinos, but it tends to seriously skew society's priorities about risk. When an adverse event with a one-in-a-million chance of occurring is popularly viewed as having a 50-50 chance of occurring, the resources spent on preventing that event are likely to be excessive. People spend far more time worrying about risks of very small magnitude—such as the most recently identified carcinogens—while engaging in quite risky behaviors—such as smoking, driving, and eating poorly—without giving them a second thought.

The disjointed public view of risk is inevitably reflected in the many laws and regulations governing risk assessment and risk management in this country. Today, at least 30 different laws regulate the use of various chemicals, including foods, in the United States. Different agencies have regulatory authority over the same risks—and sometimes apply very different standards to those risks—whereas some risks slip through the regulatory process largely untouched.

Some of the consequences of such a disjointed system are pointed out succinctly in a chapter of the budget sent by the President to Congress last February entitled "Reforming Regulation and Managing Risk-Reduction Sensibly." That chapter points out, for instance, that the cost per premature death averted by seat belt standards is about $100,000, while the same cost for the establishment of occupational exposure limits to formaldehyde is nearly $100 trillion. One cause of these disparities is that the laws creating them were based only partly—if at all—on scientific considerations.

An obvious example of a similar situation is the legislation of zero exposure to a particular substance to achieve zero risk. We still have the notice on the side of each package of saccharin to remind us of the logical inconsistencies of a strict application of the Delaney amendment. I always found the message that, for a time, appeared on soft drink cans in Canada to be at least a little more creative. It read, "It has been determined by the United States Government that the contents of this can be dangerous to the health of your rat."

In fact, the EPA recently found itself in a difficult situation on a related issue. Because of rapid increases in our ability to detect the presence of particular chemicals in foods and elsewhere, EPA discovered that many bottles of imported red wine contain detectable amounts of the carcinogen procymidone. Because of absolutist laws such as the Delaney Amendment and our inability to deal with such risks expeditiously, the FDA had to ban the import of these wines into the country for nearly a year, while interim tolerances could be established, even though EPA agrees that wine containing such small amounts of procymidone poses no measurable risks to our health—so long as we stay out of our automobiles after drinking too much of it.

The Delaney Amendment was a special focus of the President's Food Safety Plan proposed in late 1989. That plan, which was developed with input from the private sector and from all relevant government agencies, would establish a "negligible risk" for pesticide residues below which public health is not threatened, and this same concept of negligible risk could be extended much more widely. Legislation to implement the plan has been introduced in Congress, but Congress has not yet taken action.

The examples I've been discussing highlight important features of risk assessment in today's world. One is that our ability to detect and to measure the presence of particular chemical substances has increased by orders of magnitude in just the past decade—and it is going to continue to increase. Using techniques such as accelerator mass spectroscopy, chemical species can typically be detected at the parts per trillion level or less. At this level, we are going to be finding carcinogens in virtually everything, and our current regulatory system cannot deal with this situation.

Another important lesson involves the idea of risk comparison. Part of the reason for the differing approaches and conclusions in existing laws and regulations is that they were developed largely in isolation, with little effort made to coordinate their provisions. The EPA has been undergoing a very interesting exercise in comparing risks—first with the report *Unfinished Business* and more recently with the report *Reducing Risk: Setting Priorities and Strategies for Environmental Protection.* These reports provide a foundation on which the federal government can reexamine its overall approach to risk.

Finally, recent experience points to the need further to develop within the regulatory system the concept of acceptable risk. Virtually all human activities embody some level of risk. But we must bear in mind that some level of risk-taking is not only inevitable but desirable. If we were to attempt fully to eliminate all risks, it would be difficult for our society to progress technologically and scientifically. Because of our own timidity, we would condemn our descendants to less prosperous or less healthy lives than would otherwise be the case.

BRINGING SCIENCE AND GOOD SENSE TO RISK ASSESSMENT

Given the highly disjointed and confused state of risk assessment today, what grounds do we have for thinking that the situation is about to improve? Actually, I believe there are several good reasons why we should look for improvement. One of the most striking is the development of risk assessment—in our nation's universities, businesses, and government agencies—into what can be thought of as a true scientific discipline.

I also believe that federal regulatory agencies have demonstrated a more receptive attitude toward new risk assessment methodologies. I have mentioned the work going on at EPA, but a number of other federal agencies also have indicated their willingness to take a fresh look at the scientific basis underlying their approach to risk.

This new willingness to reexamine past practices is having a major effect on several activities being carried out under the Federal Coordinating Council for Science, Engineering, and Technology, or FCCSET, of which I am chairman. This is the interagency body that has the responsibility for coordinating areas of science and technology that cut across more than one federal agency. With strong support from the President, the Council has recently been revitalized and now consists of the cabinet secretaries and heads of the independent units of those federal agencies that are involved with science and technology.

Last year FCCSET established an Ad Hoc Working Group on Risk Assessment to improve and harmonize the risk assessment process in the federal government and to provide input into other policy councils in the Administration. Hank Habicht of EPA is the very effective chairman of the group, and it includes representatives from all the federal agencies involved with risk assessment.

The intent of the group is *not* to develop any manual for risk assessment or to apply specific methodologies to specific cases. Rather, it is working to develop an interagency consensus regarding fundamental assumptions and principles that can be used by all agencies to give a common base for their activities. It is looking at the different scientific methodologies used to assess risk with an eye toward reconciling disparate approaches.

Essentially, the working group is looking for ways to bring better science to the welter of laws and regulations that now govern risk assessment and management. If we can achieve some consensus about the scientific basis of risk, it will be possible to bring much greater coherence and consistency to risk assessment throughout the federal government.

Another important activity going on within FCCSET is its revitalization of the Committee on Food, Agricultural, and Forestry Research. This committee will provide a formal mechanism that federal agencies can use to exchange information and coordinate R&D programs that influence the production and distribution of food and fiber. The committee's goal is to increase the overall effectiveness and productivity of federal R&D efforts in the area of food, agriculture, and forestry research.

Recently, the committee has organized a working group to coordinate food safety research within the federal government. Thus far, the working group has identified six cabinet-level agencies that have ongoing research related to food safety. The current effort is aimed at developing a comprehensive directory of federal food safety research as a first step in identifying gaps, overlaps, and areas for interagency cooperation. It is hoped that eventually these research activities can be coordinated with those of the private sector.

PUBLIC EDUCATION

Whenever we discuss risk, it is impossible to avoid discussing a closely related issue—that of public education. Public perceptions are always going to be reflected in public policies, and that is how it should be. But we run a very real risk in having a public that is largely unfamiliar with these issues, because if citizens cannot at least appreciate the nature of the issues, quite apart from contributing to their resolution, they will tend to become alienated from the rest of society. This is a trend that no nation can long endure.

The simple facts are not encouraging. As H.W. Lewis points out in his recent book, *Technological Risk*, we live in a society where half of the American public believes in lucky numbers; less than half know that the Earth goes around the sun once a year; and American industry spends as much on remedial mathematics education each year as is spent on direct mathematics education in elementary schools, high schools, and colleges.

The facts are discouraging, but there are grounds for considerable hope. This country is now recognizing the many problems that it will face in the future if present levels of scientific illiteracy and innumeracy are not reduced. Our international competitors have demonstrated their ability to catch up to and even surpass the United States in particular areas of science and technology. Our continued technological supremacy, and with it further improvements in our standard of living and national security, hang in the balance.

The Bush Administration fully recognized the importance of science and mathematics education to our national future, and as a result it made mathematics and science a cornerstone of its effort in education. Of the six national goals that the President and the nation's governors announced in 1990, three deal directly with science and mathematics education, including the most ambitious of the six, that American students be first in the world in science and mathematics by the year 2000.

Given our current international standing, I think you would agree that this is something of a stretch goal. Yet it has already fostered some important reforms and initiatives at the state, local, and federal levels.

At the federal level, one of the most important new initiatives is an interagency program in science and mathematics education catalyzed by the Committee on Education and Human Resources under FCCSET. This committee did an inventory of everything the federal government is doing in science and mathematics education and then helped organize these efforts into a coordinated program with a long-term strategic plan.

The "American 2000" strategy released by the President in 1991 also focuses on mathematics and science as key elements of educational reform. One of the strategy's most important elements, in my estimation, is its drive to develop national performance standards in education—not a national curriculum, since this implies a degree of coercion, but standards against which performance can be measured. Such standards make it possible to give not only students but teachers, the schools, the states, and the federal government a regular report card documenting their progress in education.

In establishing these standards, I believe that, within the area of mathematics, no field is more important than that of statistics. It is essential that all our citizens be able to distinguish meaningfully and personally between an event whose occurrence probability is 0.0001% and that for which it is 1%—and neither represents a 50/50 proposition any more than does the difference between something being 90 percent safe and something being 99.9999 percent safe. We must teach our public to understand quantitative statistics and probability.

Think of the difference that a public understanding of these issues could have to the food industry. Public apprehension over the introduction of products based on biotechnology would be considerably reduced. A greater understanding of these issues could help consolidate the welter of laws and regulations that govern the activities of the food system. The ever shifting winds of public opinion might die down to at least a breeze.

The changes in mathematics and science education supported by the Administration do not just focus on the cream of the crop, on those students that have the potential to become scientists and engineers. They are directed toward all students, toward the general public that will shape future policies. I hope that I have demonstrated, with these points, why it is in the best interest of those who manage the food systems—as well as in the best interests of us all—to work with federal, state, and local governments in this effort to reform and revitalize education in this country.

DISCUSSION CHAPTER 16:

DR. MURPHY: In the setting of standards who sets the limits? Standards seem to be developed from past or present knowledge, and don't easily take into consideration the leaps of new insight that research can bring. How do we take this into account?

DR. BROMLEY: The problem is that we depend to an undue degree on media pressures in many cases. The actual setting of many of the limits takes place in the Environmental Protection Agency, in the Department of Agriculture, in the FDA. I think we have made major progress by, for the first time, getting those three agencies to actually begin to use common rules. That has happened recently. All of these agencies respond to public perception to some degree. The Alar case is a good one: science did not justify the actions taken, but media pressures led to it.

There have been earlier cases. Hexachlorophene, for example, was banned on the basis of some very shaky science done in Colorado, which happened to get onto the national news wires. The reason the ban was lifted was because my colleagues in the Yale/New Haven Hospital found that they were losing children at an appalling rate from infections that could have been taken care of easily with hexachlorophene, and so they simply decided that they would ignore the ban and go ahead and use it.

There are actions of this kind going on all the time. The thing that I think is most important is what the EPA has recognized, which is that, more and more, these decisions about limits must be based on some decent science. In the past, regulations have frequently been based on very shaky science. A number is quoted, but no measure is given of the uncertainty in that number. In my position, one of my major responsibilities is to get the President the best science and technology information that I can. But an equally important requirement is to get, at the same time, some measure of how certain we are about the particular data we are talking about. In a great many cases, we are not very certain.

DR. MURPHY: It may not be scientific as much as it is psychological, but I know when I try analyzing risk situations, I experience either unbridled optimism, confidence in my ability to solve the problem, or, under pressure, a creative bolt of lightening, which I have learned to trust. When I hear discussions of risk I say, yes, but you don't allow for that very subjective, that almost intuitive dimension. The progress that comes out of new solutions can often be limited by too strict assessment of boundaries around risk analysis. Certainly some quantification is probably useful if it isn't limiting.

DR. BROMLEY: We need the quantification because we need a basis against which to apply the sort of wisdom that comes with experience that you have just mentioned. Otherwise, the discussion tends to free float and get us into a state of chaos rather quickly.

To say that we know enough now, or will know enough in the immediate future, to give you firm numbers quantifying most of the important risks with which we deal, would simply be unrealistic. In the final analysis, we do the best we can in these formalized risk studies, and then somebody is going to apply a gut reaction to it.

DR. KORNBERG: It is appalling how bad our training is in various areas, and certainly science and math. I have recognized that it isn't the kids, of course. It is the lack of teachers, and you can't teach what you don't know. Our teachers don't know science. When on rare occasion teachers emerge who do know, they are immediately, almost invariably, attracted to a more prestigious, better paying, more secure job in industry.

I would hope we could revert to the status of teachers when I was a child. I had some excellent teachers. They were highly respected and earned a decent living. There was no condescension to accept the professional status of a teacher. Don't we have the resources to invest across the board in subsidizing the training and the employment and retention of good teachers?

DR. BROMLEY: There are three parts to an answer I would like to give you. First of all, going back to the experiences most of us remember, we have to keep in mind that, to our shame, in the days when I went to school the only two professions that really were open to very bright women were nursing and teaching. So, if you didn't want to be a nurse, you were a teacher. We as a nation benefitted enormously from the fact that we had channeled some of our brightest females into teaching.

Secondly, yes, we have the resources. We are spending more per student on education at the K through 12 level than any other nation in the world, with the possible exception of Switzerland. And we certainly aren't getting a productivity that matches that. The problem is that less than 50% of the teachers on the state payroll ever actually see students. They are involved in a large bureaucratic pyramid. We must give the classrooms back to the teachers and get rid of this middle layer of bureaucracy. That is a place we can start.

Thirdly, the committee I mentioned to you has identified, just as you have, that the serious problem is underqualified teachers in the elementary area, where only a tiny fraction have any training in the field in which they profess to teach. If they want to teach physics, chemistry, biology, the training is almost always obtained in a school of education, rather than in a normal physics, chemistry, or biology department.

We cannot wait for the next generation of teachers. We have to work with the teachers who are there now, and what we are trying to do is to use the resources of the 726 federal laboratories that we have in this country. There is a federal laboratory within a hundred miles of every school in the nation with a huge resource of people, know-how, technology, and training. These people can't tell teachers how to teach, but they can bring students in to watch actual research work being done, to motivate the students.

They can also bring the teachers in, not only on weekends and rare occasions, but throughout the summer. There is no other profession in which you expect members to go out and pump gas during the summer to maintain their income. I think we have to face up to the fact that we have to treat our teachers like professionals before we can expect them to act like professionals, and that means hiring them on a 12-month basis, and actually have them improve their qualifications during the period when they are not teaching.

It also would be a good idea to recognize that technology and the system in education froze in 1850. We still turn kids loose for two months in the summer to pick potatoes. Very few do. We have a challenge to change the structure.

DR. KORNBERG: With regard to this forum, wouldn't it be marvelous if we could enlist the private sector? We have a wealth of highly technical resources in our pharmaceutical, chemical, and food companies.

DR. BROMLEY: I am happy to be able to tell you that not only have we enlisted the private sector, getting enormous support in terms of input of ideas—that is more important than money, although money is important—it is the input of real world concepts and the kind of structure and training that an industrial employer wants among his employees. That is important to get into the system.

The other important thing is that we are bringing in the actual teachers. For example, I have hired in my own office the mathematics teacher from the Jefferson High School in Northern Virginia, who has been responsible for implementing their mathematics program, and she has been a breath of fresh air, because when people like me come up with marvelous ideas about how to improve education, there is nothing like someone who has been doing it for 20 years to tell you just how ridiculous the idea has been from the beginning. It has been a humbling but useful experience.

MR. SALQUIST: As you were talking, I couldn't help but reflect on the *USA Today* paper that I saw this morning. There was an article that the number one high school football team in the country, Banning High School in Los Angeles, had to cancel their championship football game today because of gang violence, even though the L.A. sheriff volunteered to provide air cover with armed helicopters and 200 armed policemen. I think that the atmosphere in which a lot of this teaching takes place is certainly something that has to be addressed.

DR. BROMLEY: You are absolutely right. There are a great many classrooms in the nation where self-defense takes a primary place, and that we simply should not, cannot and must not tolerate.

PROFESSOR HIGGINSON: When you refer to risk, I believe you are meaning cancer risk, since it projects most fear among the public. Of course, formalized agreements can be made between different agencies on what the risk may be in theoretical numbers, depending on assumptions. I have examined the composition of a number of bodies that have looked at this problem in the United States. It is striking that members of the society concentrating on basic cancer research, notably the American Association of Cancer Research, are singularly absent. Thus, you may find only one among fifteen other scientists. Yet, the biological knowledge on which these assumptions are made depends on the work of this group, although their society completely ignores the subject of quantitative risk assessment. This I find somewhat bizarre.

DR. BROMLEY: The point is well-taken, but you recognize that any of the groups that establish advisory bodies, the National Academies of Science and Engineering, my office, the President's office, any of these, run into a remarkable matrix of requirements for representation. It is becoming increasingly difficult to put together an advisory body in any field for any part of the Federal Government. And this is not because of more stringent laws. It is not because of anything that the Congress has done. It is simply a reflection of the fact that each generation of lawyers is more nervous than the one that preceded it.

The way the laws are interpreted makes it extraordinarily difficult. In my case, for example, I can no longer bring into my office anyone from the private sector in mid-career. I can only bring in people just out of graduate school and those who have retired. That is an enormous loss. We used to have people coming in for a year from the major industries. Then after the year, they would go back and pick up their jobs and continue.

But the new regulations require that such people, before they come into my office, have to sever permanently all their ties with their home institution. They lose their seniority. They lose their pension rights. They lose everything. You would have to be mad to do that, and, obviously, you don't want people who would be mad enough to do that in your office under any circumstances.

DR. GAULL: To change the subject a little bit, there is a distinct mistrust of science and scientists in the United States right now. Is this the Dingell phenomenon, very specifically aimed at scientists, or, as was suggested by Barbara Culliton, is it just a subset of the general anti-elitism rampant in the United States?

DR. BROMLEY: I would have to say that it is a little bit of both, and it is particularly unfortunate because it should have been a time for Americans to celebrate a remarkable period: our universities have made some unprecedented discoveries, and turned out some extraordinarily bright kids.

Instead, a combination of coincidences among a number of unrelated incidents—the Stanford situation, the Baltimore case, the South Carolina case—have cast a cloud over the universities, and there is a feeling that where there is that much smoke, there has got to be some fire. There probably is a fair amount of fraud and misconduct going on in science. There is more of that than I think there has been at any time in the past.

There is also a reaction to elitism and, quite frankly, there is a perfectly natural human reaction to some world-class arrogance that has been demonstrated in some of these recent events on the part of academics. We haven't been our own best friends in a number of these cases. I would not blame it on John Dingell.

CHAPTER 17

The New Genetics, Evolution, and Domestication: Regulatory Implications

Bernard D. Davis

In 1973 the integration of a number of esoteric techniques in molecular and microbial biology made possible the splicing of fragments of DNA from any source in the test tube, followed by multiplication (cloning) of the product in host organisms. This and related advances have had a major impact on research in virtually every branch of biology. They have also created a burgeoning biotechnology industry with many facets, including pharmaceuticals, agriculture, and bioremediation of contaminated sites.

But despite the outstanding achievements and promise of this genetic revolution, the public has been ambivalent. It is eager for the benefits, but it also fears the possibility of harm, either from accidents or from misuse. After 20 years of expanding experience with no detectable harm to man or to the environment, the anxiety has decreased. Nevertheless, the development of regulations is still plagued by confusion and controversy and by continuing public apprehension. To help explain this extraordinary persistence of concern over essentially hypothetical dangers I shall first review some unusual features of the history of the field.

THE EARLY CONCERNS: ACCIDENTAL ESCAPE

Initially the molecular biologists who created the new methodology were deeply concerned over the potential dangers. Indeed, they contributed to public anxiety by publicizing their uncertainty very responsibly—and more openly than had been customary for the initial stages of scientific discussion of yet unexplored, hypothetical risks. One reason for this unusual candor was the overwhelming novelty and magnitude of the new powers. In addition, many of the scientists no doubt felt pride, in an era of increasing egalitarianism, at being responsive to student pressure for increased public participation in the control of technology. But these scientists overestimated their ability to identify the risks, and they underestimated the public effect of their open, early discussion of the problem.

Several additional factors further distorted the framework of the discussion. The molecular biologists had created the new powers by the exercise of great intelligence and experimental skill, and they understandably viewed the assessment of the risk as a problem that they were able, and obligated, to solve in the same way. But this problem is not one in molecular genetics. It is one in evolutionary biology, epidemiology, and infectious disease. These fields were only sparsely represented at the famous conference in Asilomar, California, which laid the groundwork for the initial regulations. Moreover, at the start virtually everyone assumed—though we can now see they were incorrect—that the possible range of novel organisms was unlimited. Hence few spokespersons from the most relevant fields felt confident to make predictions about such a radically new world.

Accordingly, at the Asilomar Conference excessive caution prevailed, in a tense and rushed atmosphere in which 200 scientists attempted to thrash out the issues, in the presence of the press—quite a difference from the tradition of having a committee of experts first explore the problems created by novel discoveries and then present their evaluations to the public. At Asilomar separate working groups offered different assessments of various risks, and the one based on the most threatening scenario was recommended by the conference as the foundation for the new regulations. These were then formulated and administered by the National Institutes of Health (NIH), guided by a Recombinant DNA Advisory Committee (RAC). Because the NIH is not a regulatory agency it called the regulations "guidelines."

THE GUIDELINES AND THE RISKS

The conclusions reached at Asilomar skewed from the start the debate over genetically engineered organisms. Though the initial guidelines may have served well to meet the political need to reassure the public, they were, in retrospect, excessively restrictive. Moreover, they classified several levels of risk for different groups of organisms on a basis that pretended to be much more scientific than it really was. Within a few months scientists close to the problem realized that they had exaggerated the dangers. But it took a number of years for the NIH RAC to relax its guidelines for recombinants that proved to be harmless.

By then virtually all involved members of the scientific community had become convinced that the evaluation of a novel organism should be based entirely on its properties, and not on the technique used to create its novelty. Nevertheless, the public, and the regulators in some agencies, have not so readily abandoned the view that recombinant DNA presents a special, dangerous case. I will try to identify the main arguments that underlie this persistent dissonance, placing considerable emphasis on evolutionary principles I see as central to the issues.

First note a striking feature of this controversy: its prolonged focus on hypothetical risks, rather than on exaggeration of demonstrated ones. With newly recognized bona fide sources of harm, such as asbestos or radon, we ordinarily react slowly and then overreact after a lag. But with recombinant DNA we reacted explosively, and we continue to debate the dangers vigorously despite the lack of concrete examples.

Concern about genetic engineering has perhaps been further intensified by an underlying uneasiness over its future applications to human beings. And here there are clearly serious concerns, including the temptation to shade therapeutic purposes into eugenic ones. Moreover, knowledge of individual susceptibilities to future disease will sometimes generate more anxiety than benefit, and it will certainty create problems of privacy. These concerns have been particularly influential in those countries in which distortions of genetics had been used to justify the Holocaust: the Green political party has strongly inhibited progress in genetic engineering in Germany. Another broad concern, over environmental deterioration, has led to similar reactions in some other countries, and in Switzerland it has driven out some industries.

Still another, more general reason for uneasiness over genetic engineering has been an extrapolation from the model of the physical technologies. A few decades ago the advances in this area seemed to be providing us with a virtually free lunch; but disillusion set in as we encountered unanticipated costs for our environment, and as the release of atomic energy threatened the destruction of civilization. It was therefore tempting to argue that manipulating the cell nucleus may have just as unforeseeable costs as the manipulation of the atomic nucleus.

However, this view assumes that we have no basis for estimating future dangers from biotechnology. In fact we do. In a general sense biotechnology is not altogether novel: it has a history, which we can compare with that of the physical technologies. This history, called domestication, arose when our ancestors learned to tame certain animals, plants, and fermentation microbes to serve our needs, and then discovered how to select empirically for improved traits. The benefits have been strikingly free of social costs, for thousands of years, compared with the more mixed bag yielded by the physical technologies. Thus, the analogy between biotechnology and the harms from the physical technologies is weak.

Furthermore, despite fear that the products of the new biotechnology will "take over" by spontaneous spread, that has not been the history of the products of past domestication. They have spread, displacing the earlier occupants of the same territory, only to the extent that we have cultivated them. Since the new biotechnology is fundamentally an extension and refinement of the earlier procedures, its products should be subject to the same limitations on their spread.

These limitations arise from the nature and the scale of evolution. This process, continually experimenting with genetic novelties over the past three billion years, has been extraordinarily effective in filling each ecological niche with organisms exquisitely adapted to that environment, from the Alaskan tundra to hot vents in the depths of the ocean. The scale of this natural adaptation is particularly impressive in the microbial world: the average shovelful of soil contains as many individuals as the total human population. By comparison, our genetic experiments in the laboratory are puny.

Accordingly, the likelihood that we can further improve on the adaptation of an organism to its natural environment is virtually nil. On the contrary, breeding an organism to increase a property that serves us will *decrease* its adaptation to the environment from which the parental strain was taken.

The reason for this decrease is that the genetic changes introduced by us have costs, for they lead to a less efficient or less balanced synthesis of what the organism needs. Sometimes these costs are large enough so that the organism is dependent on our care for survival; or they may be small enough so that it can still become feral again and survive in nature (e.g., wild horses, dogs, cats). The important point is that no domesticated strain has been shown to be *better* adapted than its parental wild type to the original environment, and hence to displace the wild type there. The same can be expected for engineered variants.

There is one exception to the prediction that we cannot improve on nature: when the *environment* is changed we can sometimes predictably improve adaptation by introducing appropriate genetic changes. An example is an environment changed by the widespread use of antibiotics. This environment will select for resistant microbial strains that arise by spontaneous mutation or by genetic recombination. Under these circumstances, if we introduced genes for resistance into otherwise already well-adapted bacteria we could accelerate this shift to a resistant population. One of the aims of proper regulation of recombinant bacteria is to avoid the spread of such genes to organisms that cause disease.

LIMITS TO GENETIC NOVELTY, SPREAD, AND PATHOGENICITY

It is widely suggested that although genetic engineering has the same fundamental aims as classical domestication, we face a qualitative, unlimited unpredictable novelty! And with such a range of products some may inadvertently spread beyond our control.

Two arguments speak against this frequently expressed concern. First: even though we can manipulate DNA in the test tube at will, it does *not* follow that we can modify organisms at will. In order for an organism to develop, and also to function effectively, its parts must interact in a coordinated manner, fitting each other like the parts of a smoothly functioning machine. Hence among the variants that the new techniques will produce only those with a sufficiently coherent, balanced set of genes can survive, even under conditions of cultivation.

This argument amplifies an evolutionary principle that was just spelled out: the genetic changes introduced by us are most unlikely to improve on what evolution has produced in adapting an organism to its original environment. In general the more radical the induced changes the greater the disadvantage in the evolutionary competition. Hence recombination of ill-matched genes from distant sources will yield *poorly* adapted monstrosities, rather than the vaguely conceived dangerous monsters that current science fiction often pictures as taking over our environment.

Did these theoretical arguments from evolution and domestication carry much weight in the actual debate over risk? Probably not. I suspect that other, more pragmatic arguments contributed much more to the relaxation of the initial NIH guidelines. Among the pragmatic arguments the most obvious was the simple experience of expanding the work into thousands of laboratories, without harm.

HARMLESS MICROBES AND PATHOGENS

Finally, an additional feature of the microbial world provides further reassurance: the stringent requirements for pathogenicity. The vast majority of bacteria do not cause disease. They are found in soil and bodies of water, where they convert organic matter to simple degradation products (carbon dioxide, ammonia, etc.). This role in the cycle of life on earth is just as important as the other half of the cycle, photosynthesis. And one does not easily make an effective pathogen out of such a harmless bacterium. For with pathogens, just as we have already noted for benign microorganisms, evolutionary success is not ensured by any single, powerful gene: it depends on an effectively interacting ensemble of genes.

An example is provided by diphtheria toxin. The gene that codes for this potent toxin is found in nature only in the diphtheria bacillus, where it is accompanied by other genes that help the organism to cause disease. This toxin is not found in any members of other groups of bacteria—though it is virtually certain that such hybrids must have arisen in nature from time to time.

INTENTIONAL INTRODUCTION OF RECOMBINANTS TO THE ENVIRONMENT

With increasing recognition of these arguments against a special danger in recombinant bacteria, the guidelines for the use of such organisms in research and in industry were progressively relaxed. The issue seemed to be pretty well settled. But in 1984 a second wave of concern arose over the deliberate introduction of useful engineered organisms into the environment. Examples include their use to replace nitrogen in fertilizer, to replace toxic chemical pesticides, to digest toxic organic pollutants (such as oil spills), or to prevent frost damage on crops.

Two models were largely responsible for this new concern. One is the damage to the environment from toxic chemicals. The other is the harmful spread of certain "exotic" organisms after their importation from a distant region.

With chemicals the damage to the environment depends on the *scale* of the introduction, and it may seem intuitively obvious that the same will be true with microbes. But there is an important difference. With chemicals the harm is created directly by the introduced material, but with bacteria the mechanism would be quite different: the harm would depend on the uncontrolled multiplication of the progeny. Such spread in turn would depend on the ability of the introduced organism to compete, in a Darwinian world, with those organisms that are already present, and so we must examine the effect of scale on that competition.

In fact, if the organism is not competitive, even huge numbers can have only a transient and local effect before dying out. Conversely, if an engineered soil organism should be more competitive than the native organisms (though that would not be expected, for reasons presented above), even an accidental small escape from the laboratory could start a spread, just as a single import of smallpox can start an epidemic in a susceptible population. Scale of introduction is thus not decisive for competing bacteria. Of course, one can imagine that on a huge scale the "transient and local" effect could be significant, even though it would not result in spread. But that problem would not come as a surprise, and we should be able to control it.

The second argument for danger, the analogy to "exotics," has had widespread appeal. The unexpected excessive spread of such imports as starlings and kudzu vine in America, or rabbits in Australia, has legitimately caused great concern among ecologists. But the parallel to engineered organisms is weak, and perhaps even specious, because of a fundamental difference: one process moves an unchanged organism to a new environment, whereas the other changes the organism and then returns it to the original environment. This difference has large consequences.

The nonengineered exotic transplants have already been well adapted by evolution to their native environment, where their population density has been limited by various physical and biological factors. If these factors are lacking in the new environment the organisms will proliferate excessively. Engineered organisms, in contrast, are ordinarily returned to the *original* environment—and as we have already noted, they are highly likely to be *less* well adapted than the parental strain to that environment. Recognizing these considerations, many ecologists have withdrawn their emphasis on the model of transplanted species. However, its public appeal seems to persist.

There is still another source of reassurance: we already have extensive experience with the use of genetically modified microbes in agriculture. Such organisms, obtained by classical genetic methods, were introduced long before recombinants became available. They include *Bacillus thuringiensis* (so-called BT), used to kill insect larvae, and certain other bacteria used to replace nitrogen from fertilizer by fixing it from the atmosphere. Their regulation, to ensure safety, was straightforward, and no harm has been detected. Clearly, commercial use of recombinant bacteria, as of any other bacteria, will require similar regulatory measures, but the purpose will be primarily to avoid toxicity for humans and animals, rather than to anticipate uncontrollable, harmful spread. A parallel experience in medicine has been the use of live, attenuated viruses as vaccines, including smallpox, poliovirus, mumps, measles, and rubella.

THE REGULATORY AGENCIES

The regulatory agencies in the United States have responded in quite disparate ways to the set of conundrums created by the new genetics. A watershed, in February 1990, was the Report on National Biotechnology Policy by the President's Council on Competitiveness, a subcommittee of the President's cabinet. This report recommended a policy "to eliminate unneeded regulatory burdens for all phases of developing new biotechnology products—laboratory and field experiments, product development, and eventually sale and use." At least in theory, then, U.S. policy adheres to the principle that the government should exert oversight only to the extent that it is necessary and sufficient.

The report further notes that "existing regulatory structures for plants, animals, pharmaceuticals, chemicals and toxic substances provide an adequate framework for the regulation of biotechnology in those instances where private markets fail to provide adequate incentives to avoid unreasonable risks to health and the environment." In other words, even though it is somewhat artificial to view living organisms as chemical substances in order to apply existing legislation, the Council tacitly recognized that new legislation would be likely to cause even greater problems by treating recombinants as a special case.

The report further echoed earlier recommendations in reports from the National Academy of Sciences and its National Research Council: regulation should focus on the characteristics and risks of each biotechnology product, and not on the process by which it is created. Moreover, with a view to the future, "Regulatory programs should be designed to accommodate the rapid advances in biotechnology. Performance-based standards are, therefore, generally preferred over design standards." These principles were reinforced and further expanded in a report of the President's Council in February, 1992, recommending in greater detail the "scope" of regulation.

Nevertheless, the requirements of various U.S. regulatory agencies remain, after years of involvement, in various stages of readiness—and of compatibility with stated policy. The NIH and the Food and Drug Administration (FDA) have been closest to that policy. However, the NIH Recombinant DNA Advisory Committee has by now essentially withdrawn from all areas of genetic engineering except gene therapy in humans.

The FDA has not imposed any new procedures or requirements for products made with the new biotechnology. Moreover, in a policy statement in May 1992, on regulation of foods from new varieties of plants, it emphasized that its regulations would be based on the objective characteristics of the food, and not on the use of particular genetic techniques. It also anticipated that novel varieties could safely be exempted from regulation, with the exception of certain specified classes, such as those introducing allergens not ordinarily associated with that crop plant.

The U.S. Department of Agriculture has been somewhat ambivalent. Policies at its Animal and Plant Health Inspection Service have remained essentially unchanged since 1987—though the federal policy on "scope" would seem to dictate some refinement. Another part of the Department is engaged in developing regulations for research; current drafts seem incompatible with the federal "scope" policy.

The Environmental Protection Agency has been especially influenced by apprehensive, risk-adverse perceptions. Since 1984 the EPA has made proposal after proposal that focuses specifically on organisms obtained by the recombinant DNA techniques. All of these drafts have been incompatible with the key principles of the Council on Competitiveness.

A Federal Coordinating Committee on Science and Technology is supposed to integrate the policies of the different agencies, but its efforts, and ad hoc discussions, have not been notably effective. Moreover, problems have arisen not only from differences in the rules of various agencies but also from overlaps in their jurisdiction.

Unfortunately, large corporations, which might be expected to be most influential in this area, in general have not vigorously opposed regulations that involve extensive bureaucratic procedures—perhaps because these companies are bothered less by expense, which can be passed on, than by unpredictability, which makes planning difficult. In any case, the outcome has been hard on academic researchers and on small companies, which are often the most imaginative sources of novel products.

The apprehensive public perceptions that underlie these problems have several sources. Past costs and errors in the exploitation of new technologies have generated a cadre of political activists who are deeply suspicious of further technological developments and have become skillful in using the courts and the media to obstruct or slow such developments. Environmentalists are extremely conservative about changes in the environment, and the professional concerns of ecologists also encourage conservatism. Because the EPA has to focus on the environment it has understandably relied heavily on ecologists as advisors, and their professional society (The Ecological Society of America) has formally recommended that no proposed introduction of recombinants should be approved until it has been examined by their profession.

Unfortunately, ecologists have not been able to come up with sharply defined criteria for estimating dangers, and meanwhile years pass without evidence of harm. The reluctance of this profession to specify exempt classes of organisms therefore becomes increasingly difficult to defend. Moreover, some ecologists have advanced a curious and perhaps even meaningless argument: if we are dealing with the product of an infinitesimal risk times a potential infinite catastrophe we must be very cautious. Fortunately, such arguments, and the resulting tensions between ecologists and microbiologists or molecular biologists, seem to be fading.

CONCLUSION

The fear of hypothetical harm from engineered microbes has scant scientific foundation, except for definable classes—such as pathogenic microorganisms, which must be assessed carefully unless they have undergone changes that irreversibly attenuate their virulence. Unfortunately, it is easy to arouse suspicion of microbes in the public, since these organisms are most familiar as "germs" that cause disease. The age of biotechnology therefore now calls for a great deal of education, in our schools and to the adult public, on the beneficent, essential role of most microbes in the world. Only a tiny fraction of all microbes are pathogens, attacking organic matter while it is alive instead of waiting until it is dead.

In this essay I have emphasized evolutionary principles, as the broadest base for prediction in biology. But this reliance on theory is unlikely to be fully accepted even by some biologists—let alone by a public that is skeptical about evolution. Yet if we ignore evolutionary principles in assessing the risks, and if we try instead to rely only on empirical tests for survival or spread of the organisms under various conditions, we encounter other limitations. For we cannot hope to duplicate in our experiments all the conditions that might be encountered in nature, or all the variants that might arise.

Instead, as in all of science, we should feel confident in building on general principles, along with the examination of specific situations. I believe that for most classes of bacterial recombinants of current interest we have already examined a reasonable number of sets of organisms and conditions. Nevertheless, recombinants are still subject to special regulation, in the futile and expensive quest for the virtually absolute safety that the public has been led to expect. The cost has included not only loss of substantial benefits from products, and a discouraging impact on the investigators and entrepreneurs. In some cases there is even a decrease in safety, for example, impeding the use of engineered organisms to replace more toxic chemicals.

As a microbiologist I have dwelt on the problems of engineered bacteria, but it seems likely that engineered plants will ultimately have a much greater variety and a greater economic and social impact, as has already been seen with plants modified by classical genetic methods. The main concern with engineered plants has been that movable genes will spread from them and will create novel, harmful weeds. But principles similar to those described here apply also to this risk. It would require that several barriers be overcome: cross-pollination to a wild relative, survival and germination of the resulting hybrid embryo, fertility of the resulting plant, its competitive survival and establishment in the environment, and its creation of environmental problems. The many recombinants that have been examined have all failed to overcome one or more of these barriers to spread.

As time passes without visible harm it seems inevitable that anxiety will continue to abate, and so the regulations will eventually become more uniform and sensible. But meanwhile the success of demagogic appeals to public anxiety about the unknown has been discouraging, leading some administrative agencies to be more vigorous in responding to "perceptions of public perceptions" than in trying to help the public understand the scientific evidence. The resulting delays in testing have been especially burdensome for those scientists whose momentum depends on tests of their products in the field.

A conspicuous example of counterproductive delay has involved an ingenious use of a recombinant bacterium to decrease frost damage to certain crops. This organism, a derivative of *Pseudomonas syringae* called "ice-minus", has been *deprived* of a gene whose product initiates the formation of damaging ice crystals on the plant leaves. The organism is thus quite comparable to the attenuated pathogens used as vaccines in man. Moreover, similar "ice-minus" derivatives are found quite widely in nature, though the ice-forming strain predominates. Hence the "ice-minus" organism, used to displace its "ice-plus" parent, is as safe a recombinant as one can imagine.

Nevertheless, activists, encouraged by Jeremy Rifkin, stirred up an innocent public that lived in the neighborhood of the proposed tests, and their protests caused the tests to be forbidden successively in two of the seasons that provided the brief opportunity for testing at the required ambient temperatures. When the tests were finally conducted, with stringent precautions, knowledgeable readers could be either disgusted or amused by the picture of the experimenters forced to wear "moon-suits" while spraying the plants—a few feet from photographers plying their trade without protection from the presumptively dangerous spray.

This history illustrates the difficulties encountered when participatory democracy is stretched so far—not only brought into the formulation of policy decisions about conflicting goals, as is proper, but also, with less competence, brought into technical analyses of the underlying risks and benefits. In the problems presented by biotechnology, the courts and administrative agencies have often shared or responded to scientifically unsound, distorted public perceptions. However, with increasing experience and familiarity they will no doubt learn to apply common sense to problems of genetic engineering that no longer appear to be special.

The author is grateful to Henry I. Miller for very helpful suggestions.

SUGGESTED READING

Davis, B.D. (Ed.) 1991. *The Genetic Revolution: Scientific Prospects and Public Perceptions*. Baltimore, Johns Hopkins University Press. An analysis, aimed at a general as well as at a professional audience, by experts both on the scientific and on the policy aspects of the issues. The essay by Allan Campbell, on the history of the controversy over recombinant bacteria, is especially pertinent.

Davis, B.D. 1989. Evolutionary principles and the regulation of engineered bacteria. *Genome* 31:864-869. (Eighth International Congress on Genetics). A more detailed exposition of the evolutionary arguments presented in this paper.

Davis, B.D. 1986. *Storm Over Biology: Essays on Science, Sentiment, and Public Policy*. Buffalo, Prometheus Books. This set of essays, on various aspects of science and society, includes a number of pieces on the earlier phases of the genetic engineering controversy.

Jackson, D.A. and S.P. Stich, (Eds.) 1979. *The Recombinant DNA Debate.* Englewood Cliffs, NJ, Prentice-Hall. An excellent set of essays by early participants in the controversy, with a wide range of views.

Krimsky, S. 1982. Genetic Alchemy, Cambridge, MA, MIT Press. A detailed social history of the recombinant DNA controversy, by a social scientist with a high index of suspicion of technology.

Nossal, G.J.V. 1985, 1991. *Reshaping Life.* Cambridge University Press. An excellent survey, for a general audience, of the present state of genetic engineering.

Rifkin, J. 1991. *Biosphere Politics.* New York, Crown. A book by a leading activist, in which the introductory chapters evidence a considerable growth in sense of responsibility and concern for accuracy, compared with earlier books by the same author. In the end, however, the suspicion of technology is not substantially abated.

CHAPTER 18

Adam's Apple: A Public Policy Panel on Food Safety

Moderator: Charles R. Nesson

Editor's note: The virtually direct transcript of a 90-minute panel, including both panelist and subsequent audience participation, is provided in its entirely. Professor Nesson, working with a hypothetical worked out in advance and using his full panoply of lawyerly skills, got each panelist to play his accustomed role. There was a consensus that this exercise encapsulated, and vividly dramatized, most of the issues of the Ceres Conference. Although totally unrehearsed, it constitutes a fitting coda. G.E.G.

PROFESSOR NESSON: Let me just set the scene for you. We are dealing with the Hesperides [Editors' note: Greek mythological nymphs who guarded the golden apple] Corporation, a Fortune 500 company of which you, Mr. Salquist, are the CEO. Your scientists have come up with an apple product, the BT apple. Would you just explain to us what the product is?

MR. SALQUIST: Basically, we have taken an apple tree, ground it up into its individual cells, and transferred a gene from a naturally occurring microbe, the *Bacillus thuringensius,* or BT, and placed it into these cells, and then regenerated an apple tree. Each cell of the tree then expresses this gene, which creates a protein. When the number one pest of the apple crop takes bite of it, it dies. This product is responsive to the figures that Dr. Bromley cites: 80% of the Americans are concerned about pesticide residues in their food. This is a pest control product that is a naturally occurring protein. So, we think we have a product that will provide better apples to the consumer and make them feel better about their health.

PROFESSOR NESSON: Sounds great. The number one pest is the worm in the apple?

MR. SALQUIST: That is correct. It comes from a moth.

PROFESSOR NESSON: The coddling moth larvae.

MR. SALQUIST: That is correct.

PROFESSOR NESSON: And this is going to kill the worm because you actually have the BT protein right there in the apple?

MR. SALQUIST: Right there in the apple. It paralyzes the insect gut. When that worm takes a bite, it can't eat anymore.

PROFESSOR NESSON: Sounds terrific. Any dangers to this product?

MR. SALQUIST: No, sir. We have done several years of field trials and toxicology testing. We think this is about the safest product that is around in terms of killing insects.

PROFESSOR NESSON: In fact, the EPA has been familiar with the BT as a pesticide for some time. Yes?

MR. SALQUIST: It has been registered, I believe, for over 20 years now as a spray.

PROFESSOR NESSON: All right. Now, Mr. Salquist, you have this product. You are in the very early stages. What I want to create is the earliest strategy meeting on how you are going to put this product out to the market. How you are going to put it through the regulatory process, get it accepted, get it out there? For our purposes, I would like you to consider a strategy meeting in which you have some advisers present. You have Dr. Charles Arntzen, your agricultural adviser. You have Dr. Arthur Kornberg, a molecular biologist adviser. You have Mr. William Ruder, your public relations adviser. Talk to them. Let's formulate your strategy.

MR. SALQUIST: Well, before we start the meeting, gentlemen, I am especially pleased that Mr. Ruder is here. I had to fire the original PR agency that selected the stirring name "BT Apple." So, I am pleased to have you on board, Bill.

MR. RUDER: My first recommendation was to change it to Adam's Apple.

MR. SALQUIST: What we need to convey is the safety, a reason why consumers are going to want this product, and why farmers are going to want to grow it. I think that with the research that we have done, it is a procedural issue. Unfortunately, at this point, we must convince not just the EPA, but the FDA and the USDA. All three agencies are going to have to review this.

But the real issue is how we position this to the consumer, to convey the benefits that the product has to him, and how we deal, also, with the distribution chain, in light of the BST experience. We don't need any Ben and Jerry's out there when it comes to handling our apples.

PROFESSOR NESSON: What do you say to him, Dr. Arntzen?

DR. ARNTZEN: The first thing we have to worry about is getting this Adam's Apple into production. We have to deal with the farming community, because they are going to be the ones that grow it. They are going to want to know not so much about the nature of the BT or the basis of Adam's Apple, but how well it has been tested and whether it is going to perform—whether it is going to yield as well as other apples, whether it will store well, etc.

So, we have to get out and point out to them that this has been a very subtle change.

We haven't done anything that would ruin the traditional characteristics of the apple, but we have done something that is going to limit their costs in terms of pesticide applications and give them a more marketable product.

PROFESSOR NESSON: So, you think the number one problem, as far as you are concerned, is getting it across to the farmers?

DR. ARNTZEN: Not the number one problem, but certainly one of the first issues that we have to face, because we have to make sure we have a product that is ready to go to the market.

PROFESSOR NESSON: The farmer spends some $50 million on insecticides to deal with mites and pests with the apple. The coddling moth is their number one enemy. We have a product here that is going to deal with that. Does it sound attractive?

DR. ARNTZEN: I think it is always going to sound attractive to the farmer to find something that saves costs, but he is going to be looking at anything else that might be out there. We have to remind him of the fact that this is a new control mechanism, and we have to go through some of the data that now exists to show that he is not going to get moths that are resistant to the new insecticidal mechanism. We will have to show him or her a variety of background tests that have shown that this product is effective.

PROFESSOR NESSON: Dr. Kornberg, what is your advice to Mr. Salquist?

DR. KORNBERG: Adam got into trouble with the apple and we might also, by putting something novel in it. We have to convince people that they are buying something that is better for them and that we are not tampering with anything that is going to affect their health.

I would emphasize what my colleague has described, a very serious, methodical, scientific review that has covered all concerns about possible toxic effects and, to the best of scientific knowledge, passes the most rigid tests.

PROFESSOR NESSON: Mr. Salquist, do you have any questions for Dr. Kornberg here?

MR. SALQUIST: No. I have a high level of confidence that his team has, in fact, done the analysis that is going to meet the scrutiny of the agencies under this review and so I am quite confident that he has what is required on the scientific side to support the marketing effort.

PROFESSOR NESSON: All right. Mr. Ruder, your turn.

MR. RUDER: With the way things work now, we ought to have an approach that tells everybody everything we know, tells it all at once, and tries not to leave any kinds of open questions.

The farmer is part of the general public that gets its information or misinformation the same way everybody else does. So, you would start off by going to the total public as openly and as revealingly and, incidentally, as responsively as you possibly can, in terms of question asking and answering.

At the same time, you would specialize your communications to the farmer, to assure him that he is not dealing with a boomerang; that he is not getting into a situation where there is a potential huge turnaround. He needs the security of knowing that once he has invested considerable money and time, he is not going to find a balloon that doesn't fly.

MR. SALQUIST: I have two problems, though, that we have to deal with. One, I just spent 25 million bucks to do the science, and I don't have a whole lot of money left in the budget to do a national ad campaign and, secondly, if I do it before the FDA approves this thing, I am not so sure that I am going to speed the approval process on this. So I think we had better be pretty clever about how we are going to do this "education" campaign.

PROFESSOR NESSON: Mr. Salquist, you have some real problems with getting this through the regulatory agencies, don't you?

MR. SALQUIST: Well, there is somewhat of a state of disarray in Washington as to who is going to do what first in this case. So, it is not simple.

PROFESSOR NESSON: Have you assured yourself that this product has no risk whatsoever?

MR. SALQUIST: I have been eating them everyday for three months.

PROFESSOR NESSON: So, you have already been growing these trees?

MR. SALQUIST: You can't get the data unless you grow them out in the field.

PROFESSOR NESSON: And you figure if you have been eating them for three months, that is good enough?

MR. SALQUIST: Well, that is the tip of the iceberg of what we have been doing, but absolutely.

PROFESSOR NESSON: So, let me just put it to you again. You are telling me there is no risk with this product?

MR. SALQUIST: Correct.

PROFESSOR NESSON: This is a 100% safe product?

MR. SALQUIST: This is a hundred percent safe product with zero risk.

PROFESSOR NESSON: Is this your PR position?

MR. SALQUIST: That is my personal, scientific, corporate position.

PROFESSOR NESSON: I recently heard the President's Science Advisor telling me that there is no such thing as a hundred percent safe. He just doesn't know your business.

MR. SALQUIST: He just doesn't know my business. That is right.

PROFESSOR NESSON: I see. Now, talk to me about another kind of risk, Mr. Salquist. What is the risk of this product getting in trouble as you go through the regulatory process?

MR. SALQUIST: Well, I would say it is certainly finite. Just because my science team cannot define any risk doesn't mean that people won't perceive risk. I think science is science, but interpretation of that science, when it comes to policy, is an area for dispute. We may find ourselves getting involved in turf battles, unrelated to the characteristics of the product itself, and we may well find ourselves being attacked by philosophical opponents of technology in the food system right from the start, irrespective of the merits of the case.

PROFESSOR NESSON: Congressman Brown, if you are a fly on the wall listening to Mr. Salquist and his discussion with his advisers here, what reaction do you have to his attitude, his approach?

CONGRESSMAN BROWN: I am cynical. He is just doing what his company's interests dictate.

PROFESSOR NESSON: So, tell me about your being cynical.

CONGRESSMAN BROWN: Well, he has got a product which has got to go through some extensive regulatory proceedings, and he anticipates, with confidence, that it will. I am not nearly as convinced that it will survive the regulatory process.

PROFESSOR NESSON: Talk to Mr. Salquist for me, would you? Just explain it to him.

CONGRESSMAN BROWN: Just where do you expect to apply for regulatory approval for your product?

MR. SALQUIST: Well, we will apply at all three agencies simultaneously, but believe that the EPA, since TOSCA and FIFRA would cover this, would be the lead agency.

CONGRESSMAN BROWN: The EPA doesn't have any jurisdiction over whole plants. That is the jurisdiction of the Department of Agriculture.

MR. SALQUIST: Well, in terms of growing them, it is the USDA. In terms of eating them, I believe that the FDA has the authority, and in terms of killing bugs with them, I think that the EPA has the authority.

CONGRESSMAN BROWN: This illustrates one of your problems, doesn't it? I am concerned about whether or not the regulatory process is adequate to deal with the product that you are offering here.

MR. SALQUIST: Well, I am concerned also, Congressman, but I am not concerned with the technical capabilities of the people in the agencies to deal with this. I mean, I think that there is the toxicology expertise, the allergenicity expertise, you name it. I think we are really dealing with an administrative morass here, that too many cooks spoil the soup.

CONGRESSMAN BROWN: How did you get those apples you say you have been eating? Did you get permission from the Department of Agriculture for a field test?

MR. SALQUIST: Absolutely.

CONGRESSMAN BROWN: And have they been following the course of the development of the trees and the fruit?

MR. SALQUIST: That is correct. We have to report every time we do a field trial and we are in our third year of trials now and they have —

CONGRESSMAN BROWN: Everything has gone okay so far?

MR. SALQUIST: No unforeseen circumstances.

CONGRESSMAN BROWN: Have you had any problems with the possibility that there would be resistance developed to the BT?

MR. SALQUIST: To BT resistance? Well, this is an area where I would say there is more scientific divergence of opinion than others. However, the issue of the build up of pest resistance to a continual exposure of BT, instead of spraying it on and periodic exposure, certainly is a question that has been raised. I would defer to my agriculture adviser, Dr. Arntzen, to respond to what the possibilities of that are and how we have evaluated them.

PROFESSOR NESSON: Mr. Salquist, let me ask you a question. Congressman Brown here suggests that maybe the road is going to be a little rocky. He is skeptical of the future and you in your description haven't really worried out loud about what Jeremy Rifkin might do or what Roger Blobaum might do or what Mr. Ubell might do or Ms. Culliton might do. I mean, your picture seems to be of a nice, organized, scientific administrative process that you are going to go through.

MR. SALQUIST: Well, that aspect of the process is. You have just turned another page, however, and there is this vast array of external pressures on the agencies and in the public arena that obviously has to be addressed.

PROFESSOR NESSON: So, my question to you is: Is it worth it? I mean, look at this product. What are you going to get out of this product?

MR. SALQUIST: Well, first of all, this is just the first of a whole series of products that we will be introducing in the fruit and vegetable industry. The apple happens to be the most attractive target for us because we think there is a compelling demand for it at the consumer side, and it will pave the way for a whole series of other products.

PROFESSOR NESSON: I see. There is $50 million total in the pesticide application for apples. Maybe $15 to get coddling —

MR. SALQUIST: Well, we are not in the pesticide business. I am not going to make a dime off of that, but there are a billion dollars worth of apples consumed. I am going to make my money selling apples.

DR. KORNBERG: Aren't we getting a little diverted from the importance of bringing out better products that are safe for the consumer, rather than worrying about whether my CEO is going to make money?

PROFESSOR NESSON: I see. The purpose of his corporation is to serve the public interest. Money is unimportant.

DR. KORNBERG: No, it is not unimportant.

PROFESSOR NESSON: He is in the business to bring out better products.

DR. KORNBERG: Why impugn his motives unnecessarily? He does want to produce a better product, through which he will have a better business.

PROFESSOR NESSON: I didn't think I was impugning his motive. Did you consider your motive to be impugned, the idea that you need to make a profit?

MR. SALQUIST: Well, I would love to be accused of making a profit.

DR. ARNTZEN: I think you are also getting off the track because you are focussing on one little aspect of this new apple. We introduce new apple varieties almost every year. Through the breeding process we are constantly improving the quality of the apples in terms of taste, storage characteristics, color, etc. All of these include a number of chemical and constituent changes. This is, as I see it, just one more step in a logical continuation of improvements.

PROFESSOR NESSON: So, we may not even need to worry about all this regulatory hassle. This is just a new apple. That is all. This is just like we grafted one variety to another variety. We come up with a new variety of apple. That happens all the time. What is all this worry that the Congressman has?

DR. ARNTZEN: Because of the way in which the apple was produced, using recombinant gene technology. It therefore must come under the guidelines that have been established and that the industry accepts.

PROFESSOR NESSON: You mean it makes a difference who made this apple? I am eating an apple. I have no idea as I am eating the apple what produced this apple. For all I know, it was made by little green giants.

DR. ARNTZEN: In a regulatory sense, it makes a difference. In terms of the biological components, it does not make any difference.

PROFESSOR NESSON: If he could have produced this apple by grafting plants together, there would be no problem?

DR. ARNTZEN: It would not have come under regulatory supervision.

MS. CULLITON: The fact is, though, that it does come under regulatory supervision, and I think that Mr. Salquist himself gave one bit of evidence as to why people might be concerned about food safety. It is the question of whether or not the recombinant technology could lead to BT resistance. Now, that is not an issue that comes up if you are just grafting one plant onto another.

And he acknowledged that there is some difference of opinion within the scientific community about that issue. Therefore, I would like to ask him whether or not he would be willing to publish his research data in the open scientific literature, rather than holding it as proprietary information, and thereby allow everybody who is technically qualified, as well as members of the press, to look at his evidence of safety and see what they think about it.

PROFESSOR NESSON: Mr. Salquist, how are you going to announce this product? The world doesn't know about it. Ms. Culliton doesn't know about it. The Congressman doesn't know—nobody knows about this product. How are you going to announce it?

MR. SALQUIST: Well, first of all, the way the biotechnology industry works, the day we filed for a field trial application with the U.S. Department of Agriculture, it became public knowledge and was accessible to anybody under the FOI, and it was published in *Food Chemical News* and a number of other publications.

PROFESSOR NESSON: So, that is the way we are going to announce it?

MR. SALQUIST: We put out a press release as well. I am not totally oblivious of my stock price, and every time one makes a tangible advance in the biotechnology world, the market tends to respond. So, it is to my best interest, if I have a good product, to have it publicized as widely as possible. And we will make available, to anybody who wants it, the complete data filing package on this.

PROFESSOR NESSON: So, your answer to Ms. Culliton is "yes," just come right in.

MR. SALQUIST: Absolutely. Give me your card.

MS. CULLITON: You would be willing to publish the data in a scientific paper subject to peer review by other scientists?

MR. SALQUIST: Data that is required to meet FDA documents is not always of sufficient interest to get peer review.

PROFESSOR NESSON: Mr. Salquist, your adviser here is very hot to answer this question.

DR. KORNBERG: Early on, as you recall, we published the early results regarding this bacillus gene product. We published it promptly. It appeared in *Cell* and in *Nature*. It drew a lot of attention and our scientists are free to communicate what they know. And as you have pointed out, it is covered by patent at an early stage, but there has been no withholding of any of the basic scientific information. All the properties of this fruit and its toxicology have been properly reported in the open literature.

PROFESSOR NESSON: Mr. Blobaum, you pick up the *Food Chemical News* and you read about Mr. Salquist's applying to the FDA for field tests of the BT apple. What is your reaction? There you are over breakfast with your coffee.

MR. BLOBAUM: Well, I am surprised. First of all, I would suggest that Mr. Salquist set aside a little money for Mr. Ruder. I think that failing to allocate any money for public relations would be a mistake.

PROFESSOR NESSON: Why a mistake?

MR. BLOBAUM: Because I think he is going to need all the public relations help he can get. I think he made the wrong choice by selecting the apple as the fruit he would start with. It picks up, from a consumer standpoint, all of the bad reaction that surrounds Alar. It raises everybody's suspicion. It is just a really bad choice. I think that he is going to have to polish the apple a little bit with consumers, right off the bat.

I also think that Mr. Salquist would have helped himself a great deal if he had at a very early stage come by to brief me and many of my colleagues in the organizations that are going to respond to this. He should have told us what he was up to and shared with us what information he had. I don't like to read about important developments like this in the paper.

PROFESSOR NESSON: What is the public office that you hold?

MR. BLOBAUM: I don't hold any public office.

PROFESSOR NESSON: Why is he coming to see you?

MR. BLOBAUM: Because I am a player in this. I am part of the food system also, and the environmental and consumer organizations are a legitimate part of the food system, and we have a legitimate role to play and, in fact, Mr. Salquist's whole process will be enriched by our activities.

PROFESSOR NESSON: Mr. Salquist, do you want to talk to him?

MR. SALQUIST: Well, that is obviously on our agenda, Roger, and we have lots of bases to cover, and I don't disagree with your comments at all.

PROFESSOR NESSON: Would you talk to him? Would you call him up? He is saying before I read about this in the *Federal News*, I would like to have a call from Salquist.

MR. SALQUIST: I don't have any problem with that.

PROFESSOR NESSON: Would you have Ruder talk to him?

MR. BLOBAUM: No, I don't want to talk to Ruder.

PROFESSOR NESSON: He is the guy you hired to do PR.

MR. SALQUIST: I would have Dr. Arthur Kornberg and Dr. Charles Arntzen talk to him. You bet.

PROFESSOR NESSON: Whom do you want to talk to, Mr. Blobaum?

MR. BLOBAUM: I don't want to talk to Mr. Ruder. I want to talk to Mr. Salquist.

PROFESSOR NESSON: Call him up, Mr. Salquist, and say "Hello, Roger."

MR. SALQUIST: Well, Roger, we have this apple coming down the line and I know you are concerned about this area. So, I am more than happy to buy lunch for you next week. Let's sit down and hear what your concerns are specifically.

MR. BLOBAUM: I am not necessarily opposing your apple, but I do have some questions that I would like to pose because I think they would help you in the whole process, as well as help me when I start getting calls from our members, who read about this in the paper.

MR. SALQUIST: Yes. I think that is a legitimate point and I am more than willing to get together with you at your earliest convenience.

PROFESSOR NESSON: I am actually very interested in this, Mr. Blobaum. Your feeling is one of affront that he didn't call you?

MR. BLOBAUM: I think that is true.

PROFESSOR NESSON: Should he have a list of everybody who runs a public interest organization that he ought to call, everybody who is in the food system he ought to call? I mean, why does this guy have to call you?

MR. BLOBAUM: There is a tendency on the part of industry people not to consider environmental and consumer organizations as part of the food system. I insist that we are a legitimate part of the food system with a responsible role to play, and I intend to play that role. I am going to be involved in this whether Mr. Salquist calls me or not. That is the point. And so are my colleagues. I am just giving him some good advice.

PROFESSOR NESSON: All right. So, fill us in, Mr. Blobaum. You have picked it up. You have read about it in the *Federal News*. You are affronted that this guy hasn't called you. I guess he is new in the business, doesn't realize that is what he was supposed to have done.

What is your strategy? What do you do now in response to this new information that the BT apple is going to go out for a field test maybe?

MR. BLOBAUM: Well, one of the main questions is to what extent there will be transparency in this process. I think that is a very important thing to establish at the beginning. Consumers and environmental groups also ask if there are other alternatives. Is the choice simply between applying a pesticide to control the coddling moth or using BT within the apple? How about putting these orchards under organic management? How about spending some money to make organic management of orchards more effective? How about more use of IPM? Maybe that would deal with the coddling moth problem. I mean, I think there are other alternatives.

PROFESSOR NESSON: What don't you like about this product? This product sounds great.

MR. BLOBAUM: I think this product will be extremely difficult to sell to the American public. That is one thing.

PROFESSOR NESSON: Well, let him worry about that.

MR. BLOBAUM: Well, I worry about it, too.

PROFESSOR NESSON: Is that your public interest worry? You worry about whether his product is going to make money on the market?

MR. BLOBAUM: No, I don't worry about whether it makes money. I know that I am going to get many calls and letters and pressure from my members to oppose this project, and not necessarily to oppose it responsibly, either.

PROFESSOR NESSON: Mr. Arntzen, you talk to him, would you?

DR. ARNTZEN: I am coming into this a little bit surprised, because we have an experience base and history for new apple varieties. We publish extension bulletins on the performance of new varieties and their uses. No one has ever asked us, nor have we ever gone to, Roger's organization or any others for approval of new varieties. We are involved in agriculture 365 days a year, and this is not an unusual process—just another new apple variety.

PROFESSOR NESSON: What is he worried about? Mr. Rifkin, you get up in the morning —

MR. RIFKIN: This is not a new apple variety.

PROFESSOR NESSON: Actually, you are up at 4 o'clock in the morning. You never sleep. Here is the *Food Chemical News* and here is the announcement. What is your reaction to it?

MR. RIFKIN: This is a very interesting experiment and product that this company has come up with. I don't know whether to be amused or chagrined by this hypothetical that has been presented this morning. Some of you might have a little sense of deja vu.

PROFESSOR NESSON: This isn't hypothetical. We are in the real world here.

MR. RIFKIN: I am going to give you the real world. An experiment almost identical to this has taken place and let me tell you what the problems were. A company called Crops Genetics, Inc. wanted to introduce BT into corn in order to kill the corn borer - almost an identical experiment, different produce. They had corn instead of applies. They said it was perfectly safe. It was a start-up company. It was efficacious. There would be a market for it. The farmers would like it, and the consumers could accept it. Let me explain the problems in this product from a scientific perspective, as opposed to a marketing or public reaction perspective. The problem was not just with the BT. While BT poses some problems, as always with these experiments with recombinant DNA, what distinguishes these types of experiments from traditional breeding experiments is often the vector you have to use to introduce the gene into the organism.

PROFESSOR NESSON: Hang on for just a second, Mr. Rifkin. You read the *Federal News*. You look at this thing and I get the picture. You are saying, oh, my God, look at these guys. They are at it again. I want to know what you do.

MR. RIFKIN: I have to explain to you by telling you what the problem was, and then I will show you how we went through it.

PROFESSOR NESSON: Who are you going to explain it to?

MR. RIFKIN: I think we need to get beyond rhetoric here and get down to science and how we would go about doing what we are going to do. Give me one minute to explain about an almost identical product and how we went about it. All right? I think it would be informative for the people here, because many of the companies in this room are dealing with everything from bovine growth hormone to tomatoes.

The problem with the corn, and we consulted with scientists across the country, was this. The vector that they used to put the BT into the corn is a known plant pathogen. It happens to be a bacterium, Clavibacter, a pathogen for some 70–80 species. It causes Bermuda grass stunting disease, and it is also a pathogen towards certain domestic crops, like squash.

When the company submitted this data to the government, they found in the lab experiments that the vector with the BT in it could spread. If you cut open the seedling with the BT inside, the vector spread onto tabletops and into the environment, and during this process, scientists became concerned that the vector they were using, a plant pathogen, could spread its host range.

PROFESSOR NESSON: Actually, Mr. Rifkin, I tell you what I would like you to do. I mean, the way this hypothetical works, I would like you to talk to someone. You have a point. I want to know who you are going to make it to. You wake up in the morning. What are you going to do? Who are you going to talk to?

MR. RIFKIN: What we did in this one and what we would do in the current example on the table is this: We check with scientists around the country, primarily those in the environmental sciences, but also those in the molecular biology community, those who will talk and who are not afraid that they will lose their research grants by either confiding in us or becoming a consultant to our organization. Then we go to groups like the National Wildlife Federation, Audubon, and the Environmental Defense Fund, all of whom were involved in this other product that I cannot talk about. We consult with their scientists. After we have consulted with them, we take a look to see if there is a potential risk involved in growing this, an environmental risk. Is there a risk in marketing it to the consumer, a public health risk?

If we conclude that the experimentation to date raises more questions than it answers, and that there is enough scientific doubt, then we will petition the appropriate agency, as we did in this other case, bringing together the best science we can, in order to ask important questions that have to be answered during the review process.

If the agency refuses to address those questions with the company, or if the company refuses to provide the appropriate data that would arrest our fears and concerns, then we would have to take the next step after the petition process, which is legal action.

PROFESSOR NESSON: Hold on before we get to the next and the next step. I am still interested in your first step.

DR. KORNBERG: The case that you have presented, Mr. Rifkin, is not analogous. We are talking about oranges and apples, in effect, because the vector here is simply injecting something harmless into cells. We are not dealing with a plant pathogen. We are not going through the scenario that you have talked about. It has nothing whatever to do with our case.

PROFESSOR NESSON: Mr. Rifkin, it is still early in the morning. You have read this article. You have had your initial reaction. You have nothing back from the company. You are not in a dialogue yet. You recognize it as something you have been through before with corn.

You spend the morning on the network with all of the groups that you are familiar with, the scientists who will respond to you, and you get convinced that this is bad.

MR. RIFKIN: You are short circuiting the process. We don't become convinced it is bad in one day or one month. Many of the products that we are looking at, products here in this room that your companies are trying to bring to the public, undergo months and sometimes years of analyzing as you are moving them through the pipeline.

Mr. Salquist has a tomato that he would like to put out. We have been looking at it for a long time. We do not make split-second decisions because we are aware of our responsibility to the public and our good name. We do not raise issues unless we feel there is some scientific or public health reason to raise them. And as some of you know in this room, some of the products you have brought for testing have not been challenged. Why? Because we did not find sufficient reason to believe they would be an environmental risk or a public health risk.

We don't make a decision in one day. We carefully look at the information. If we need to, we go through FOIA. I must say that there has been a little unreality at this table this morning. I can't tell you how many times we have been denied appropriate information under the Freedom of Information Act by these agencies under the guise of proprietary trade information. The fact is that we get very little out of FOIA, and so we are forced many times to spend months and years trying to find friendly scientists who will do independent experiments or give us a lead, but it doesn't happen in one day or one month.

PROFESSOR NESSON: Fair enough. It happens with this product at some point. Assume that.

MR. RIFKIN: Yes.

PROFESSOR NESSON: I want to know what you do next. I want to know, is the first thing to file with the FDA? Or is your first thing to somehow go public with public relations, the way Mr. Salquist was doing first? What do you do?

MR. RIFKIN: It depends on the urgency of the situation. I can give you the normal procedure.

PROFESSOR NESSON: This is pretty urgent. We have it in front of the FDA. You don't know how they are—

MR. RIFKIN: Well, I would take this case and say this was not urgent, because it can go through the normal review process. So, here is what we would do.

Our first step would be to petition the agency, a legal petition, in which our attorneys draft up the scientific evidence that we have with accompanying affidavits of scientists that have worked with us. We petition the agency with our concerns, and raise questions we would like the agency to address and the company to deal with, and further data that is presented for its case for an application. That is the first step. It may take six months for the government to get back to us. We wait. We often do not go public. In fact, there are many products in this room that are under review by our organization that have never gone public by our organization.

We wait until we go through the administrative process. If the agency gets back to us—they always eventually do—and if we are satisfied with the results, and in one recent case we were, we will go no further. If we are dissatisfied, and believe that they are either stalling or forestalling, we will then take the next step. The next step either is to involve members of Congress, to initiate litigation in federal court, to go public, or perhaps to create a coalition of organizations that may try to stop any further development of the product. There is a range of institutional devices that are available for the public to become involved in the process.

PROFESSOR NESSON: Would you consider writing a little piece for *USA Today*? I have seen your name there before.

MR. RIFKIN: I would consider it. It depends on the product involved.

PROFESSOR NESSON: It is BT apple. You tell the corn story and how this has been done before and how this was —

MR. RIFKIN: You have to understand. My frustration here is that I have been involved in this process for a long time, since the first deliberate release, and I ought to write a book about the chicanery and the shenanigans that go on in federal agencies and among the so-called scientific establishment putting their imprimatur and blessing on products. The fact is there are companies in this room that have a long history of bringing out products, some of which are safe and many of which are dangerous, damaging to the environment and public health.

So, given that track record, it makes all the sense in the world for the public and public interest groups to be skeptical, perhaps cynical, and to demand participation in the review process. The companies themselves have self-interest as opposed to public interest—

PROFESSOR NESSON: Would you consider holding a public press conference?

MR. RIFKIN: Absolutely. It depends on the situation.

MR. SALQUIST: Well, I would suggest, Mr. Rifkin, that your book would have to have at least half the chapters on the chicanery and false scientific analysis from your organization. The first example would be that of the predictions that you made with respect to your analysis of the frost ban field trials, of which you were a rather outspoken proponent, not one of the dire predictions that you made came true; nor were they solidly based scientifically.

MR. RIFKIN: Well, you are absolutely wrong on that, sir. The predictions we made came from scientists at the National Oceanic and Atmospheric Research Institute, who said if you go from small field tests to a massive use of frost ban in commercial agriculture, it could potentially change the balance and rainfall patterns. Because they did not go to commercial application for a whole series of reasons we could discuss later, we weren't able to see whether that prediction would hold true.

PROFESSOR NESSON: Ms. Culliton, Mr. Rifkin holds a press conference. By the way, Mr. Blobaum, are you going to be at his press conference? You are going to hold your own press conference.

MR. BLOBAUM: Yes. What I am going to do is consult some lawyers and scientists that I know and trust and decide on a course of action. One of the things that I would certainly take a look at, which would probably disturb Mr. Salquist, would be to see if this couldn't be considered at FDA as if it were a food additive, since the BT is being added to the apple. To get that resolved, we need to establish at the outset whether products of this kind involve food additives in the legal sense. That would be a slightly different way of looking at it and I think a legitimate question to ask.

PROFESSOR NESSON: Ms. Culliton, he holds a press conference. Are you there?

MS. CULLITON: Yes.

PROFESSOR NESSON: Would it make a difference if, say, he held it on a Sunday? Would you be there on a Sunday?

MS. CULLITON: Yes, we would be at the press conference if it were terribly important. Generally, we would expect it to be on a weekday. We would be at it because of the things that Mr. Rifkin has said and the things that Mr. Blobaum has said. There are two parts of what we do. One is cover the news and the other is publish scientific data. This would be news.

PROFESSOR NESSON: Let me make you the science editor of a major newspaper. Does that put you at his press conference?

MS. CULLITON: In the same way, yes.

PROFESSOR NESSON: You hear all the statements he makes. He lays out the science. He lays out his position. He lays out the analogy to corn BT. What do you do with that?

MS. CULLITON: First, I would try to get him to give me the scientific basis for his concern about the apple, as opposed to his political concerns about technology, and then I would ask scientists I trust whether or not they trust the scientists he consulted.

So, I would ask if there is a bias one way or the other among the scientists with whom Mr. Salquist has consulted in order to demonstrate that this apple is good; is there an antiindustry or antitechnology bias in the scientists with whom Mr. Rifkin has consulted to challenge the issue? That would bring me right back to asking Mr. Salquist whether he is willing to publish all of his data.

But I wouldn't believe Mr. Rifkin's scientists on the surface anymore than I would believe the others on the surface.

PROFESSOR NESSON: So, that is what you think your job is, to figure out whom you believe?

MS. CULLITON: Yes.

PROFESSOR NESSON: If you don't believe Mr. Rifkin's scientists, their message doesn't get out?

MS. CULLITON: Their message gets out, but it gets out in a way that indicates that my newspaper finds it a less credible message than the alternative message, than the opposite message.

PROFESSOR NESSON: Write the column for me, would you? Mr. Rifkin's scientists say BT apple is bad for you. The BT in the apple is going to make the little worms die.

MS. CULLITON: I would say that Mr. Rifkin has raised again the same questions that he has been raising for 20 years about the safety of biotechnology in agriculture and the food supply. These concerns have, so far, not materialized as a substantial threat to the public health. He did not present evidence on this particular occasion to make me frightened of eating Mr. Salquist's apples. I might or might not eat them for other reasons, but I would report this as an environmental versus science conflict that is entirely consistent with the polarized positions that we have seen over the past 20 years.

A news story would say that Mr. Rifkin has indicated that this apple is likely to cause cancer or some other hazardous effect on your health, and if I were the editor of the newspaper, I wouldn't publish it that way. I would think that reporting only one side of this issue would not be responsible journalism.

PROFESSOR NESSON: What would the headline be on your story, news story?

MS. CULLITON: "BT Apple Causes Cancer, Rifkin Says."

PROFESSOR NESSON: Good. Tell me if the following statements are true. Bad news is good news. Is that right in the newspaper business?

MS. CULLITON: Oh, absolutely.

PROFESSOR NESSON: How about good news is no news?

MS. CULLITON: That is probably true also.

PROFESSOR NESSON: How about no news is bad news?

MS. CULLITON: It is bad for business.

PROFESSOR NESSON: So what it says is bad news is news. From the newspaper business point of view, Mr. Rifkin is going to get very good play with his press conference?

MS. CULLITON: Yes.

MR. RIFKIN: Let me disagree with you on this. The fact is that good news from the industry and the scientific community gets much more play in the biotechnology area than bad news. For example, every time a small experiment is conducted that may suggest a long-term cure for any one of a number of diseases, it is front page headlines without any voice from the other side cautioning or showing concern. When new products come out in a whole range of biotechnology fields, they are heralded. You see front page headlines in the *New York Times*, the *Wall Street Journal*, *USA Today*, and the *Washington Post*.

What I have witnessed in the last few years is a hardening among science reporters and science desks. The word around the public interest community is that the major newspapers and magazines are intent on only making the public literate about science and technology advances, and are covering the potential problems and cautions with less vigor. That is not to say we are left out. But I can't tell you how many times I open up the paper every week and I see some new medical, agricultural, or other breakthrough in biotechnology that is put out without any caveats or qualifications.

So, it is not true that bad news is the only news that is covered. In fact, to the credit of Ms. Culliton's journal, *Nature*, this last week we finally raised the issue of eugenics in discrimination when it comes to the technology of genetic screening and human genetic engineering. We have fought with the editor of your magazine and with the scientific establishment for years to get that out in the public. To the credit of your publication and a few others, finally last week or two weeks ago, in Congressional hearings, we were able to publicly address for the first time the problems that screening and genetic technology may raise in terms of the social and public dynamics of society. It took 14 years to get to that point.

MR. RUDER: So much of this is in the eyes of the beholder. When we are involved in a new story, we are hypercritical and tend to see only the negative sides of it. When one looks at it objectively (if one ever can), it is a different kind of an article and very often well-balanced. I would like to turn the clock back if I can. I think we have taken this exercise down the road of the worst case scenario. It was useful and productive to do that, just to demonstrate how counterproductive our knee-jerk reactions can be.

If at the outset we had decided that there were five or six different constituencies that would be interested in this, and if we had gone to some of the public interest and consumer groups, sat down with them individually, suggested to them that we might even have a group meeting among them, if Jeremy Rifkin, for instance, could have brought up his analogy to the corn situation in such a meeting, there could have been a dialogue and perhaps he might have been convinced that this is not an analogous situation.

PROFESSOR NESSON: So, you would have set up a press conference with—

MR. RUDER: No.

PROFESSOR NESSON: You want Mr. Rifkin to come to a private meeting?

MR. RUDER: Yes.

PROFESSOR NESSON: To be persuaded.

MR. RUDER: Not to be persuaded, but at least to get our information and data, and for us to understand what his problems with it might be. Also, to see if there were a way that the two of us could look together at allaying those concerns. I think we could have done it with science editors in the journalistic community. I think we could have done it with some uninvolved scientists along with our scientists.

Before you leap into this, it seems to me, you prepare the ground very carefully and very thoroughly with those people you know are going to have a particular interest in it, and (a) try to get them to understand what you are doing; (b) try to hear from them what their concerns are; and (c) see if you can't narrow the distance, if, there are distances, between the positions.

PROFESSOR NESSON: Mr. Ubell, let me make you the producer of a documentary investigative reporting vehicle.

MR. UBELL: Okay.

PROFESSOR NESSON: Are you interested in Mr. Rifkin?

MR. UBELL: Sure. I found myself agreeing with him.

PROFESSOR NESSON: Are you interested in the BT apple as a story?

MR. UBELL: Sure.

PROFESSOR NESSON: Talk to me about how you might think about putting together a documentary piece on the BT apple.

MR. UBELL: Well, first of all, I would try to judge what was really new about it and what the benefits might be and then I would try to get around to Mr. Rifkin to see what he had to say about it.

PROFESSOR NESSON: Talk to Mr. Rifkin for me, would you?

MR. UBELL: I would like to know from you if you have it, what evidence you have that this is bad or good or whatever. And we would have a conversation. What do you have?

MR. RIFKIN: I would provide any evidence we had, from documents we had gotten, that would be my first interest. I would provide concerns, and then I would put you on to some scientists that we work with, who could talk to you about their concerns either on the record or off the record; hopefully, scientists that you could interview. Because I think when you get to the science of this, it is important to have people in the field or related fields talking.

I would then say to you that there are two different perspectives or world views when it comes to this apple. One is that of the molecular biologists and the second is that of the environmental scientists. I was interested, by the way, to note that somebody on this panel said I would look at the environmental side and then the science side—which is a real clue to me. Environment has a science that goes with it and it is no less legitimate than the molecular biology.

I would tell you to find out what the molecular biologists are saying at the company and then listen to what the environmental scientists are saying, and then you will make your story any way you see fit.

MR. UBELL: Well, I would like to see the list of scientists. And we all know that scientists can take different positions on complex things, as was shown in the Alar case, where press conferences were held by scientists who thought the Alar case was egregious.

Then I would try to figure out a way in which to produce that in the context of having talked to others.

PROFESSOR NESSON: He has a lot of people who are upset about the BT apple. Are you interested in them? Are you going to interview some man on the street, woman on the street about eating this apple?

MR. UBELL: That doesn't prove anything.

PROFESSOR NESSON: Is that what you are out to do, prove something?

MR. UBELL: I am out to show what the sides are saying, what the two sides or three sides or four sides are doing about this.

PROFESSOR NESSON: Who do you want from the company side?

MR. UBELL: Dr. Kornberg, of course.

PROFESSOR NESSON: Dr. Kornberg, of course. Dr. Arntzen, do you want to say something?

DR. ARNTZEN: I am just curious. What if Roger were to hire Jane Fonda to come out and eat an Adam's Apple? Would you show up, and would she make news if she gave an endorsement?

MR. UBELL: Not to me. It is a problem. Who are the scientists that one can talk to and trust? Ms. Culliton said that she has scientists whom she trusts. So do I.

PROFESSOR NESSON: So does Mr. Salquist.

MR. UBELL: Yes, of course.

PROFESSOR NESSON: So does Mr. Rifkin.

MR. UBELL: Right, of course. And the question would be, how do you untangle this?

I would have to see what the specifics were. Who Mr. Rifkin is offering, what data he has and so on, compared to what the biotechnology house is doing.

PROFESSOR NESSON: When you go to the corporation, you get Mr. Salquist, let's say, and Salquist gives you a blast about Rifkin. What do you have to say, Mr. Salquist?

MR. SALQUIST: Well, I would say that it would be refreshing if Mr. Ubell for once would question the motives and the mandate of the people who were challenging industry in this case. I mean, Mr. Rifkin isn't in the business of protecting the public interest. He is in the business of raising money for his foundation. It would be nice to question his motives in the science of business because they are supposedly making a profit. Mr. Rifkin is in a business, too.

MR. UBELL: He would turn me off, I have got to say, Jeremy.

MR. RIFKIN: I really resent what you just said, because in all these years I have never impugned the motives of an individual in this room. If you take a look at the writings, our press conferences, our petitions, we have never ever attacked personal motivations, whether it be the genetic engineers or the scientists. We have talked about profits of the company, but not individuals.

I believe that you are in business because you want to put out a product. You want to make some money. My family is in business. I would not impugn your motivations. I am involved in this because I truly believe in what I am doing. I have spent a lifetime doing this and I think that it is a sign of disrespect when we have to treat each other that way. Let's deal with the issues at the floor and not with personalities. I wouldn't do it with you.

PROFESSOR NESSON: How does it make you feel when they call you names, Mr. Rifkin?

MR. RIFKIN: It has been tough.

PROFESSOR NESSON: You are a Luddite?

MR. RIFKIN: Well, it has been difficult. I am 46 years old. I have gone through this for a long time, and I am not insensitive to it. What I have tried to do over the years is, hopefully, set an example by not reducing myself to those kinds of invectives, and I have not done it with anyone in this room.

All I am saying is, if we want dialogue, there are legitimate differences of opinion. We ought to be able to air those and be passionate, and many times I have said I don't think this product is any good, but at least we ought to keep the dialogue on a high enough level so that the public can respect both of our positions, even if they disagree with one of them.

PROFESSOR NESSON: Mr. Salquist, what do you actually think Mr. Rifkin's motives are?

MR. SALQUIST: I think he is in the business of creating controversy and funding his particular efforts.

PROFESSOR NESSON: You don't credit him with the same kind of good faith that you have in wanting to put out a good product?

MR. SALQUIST: I have a higher level of confidence in Mr. Blobaum's approach to the issue. I think there is a spectrum of public interest people, and I think that Mr. Rifkin is at one extreme of misrepresenting the science —

PROFESSOR NESSON: You don't think Mr. Blobaum is in business to raise money for his foundation?

MR. SALQUIST: I think he is. But I think there is a spectrum of how one goes about it and how one tends to view the science.

MR. RIFKIN: I think the way one should go about it is by using all the legal avenues of the system, and saying things that are responsible. Who are the coalitions we work with? Let me give you the names. They include Audubon, National Wildlife Federation, Consumers Union, National Farmers Union—a whole range of responsible organizations that we have worked for, and coalition after coalition that we have spearheaded.

Are you saying that our motivations are to be impugned because we simply disagree with what some of these companies have done in here? We have a track record. What you have to understand here is that there is a new chapter in relation to your products. Science and technology products more intimately affect our personal and political lives sometimes than all the decisions made by the parliaments of the world; yet, up until recently, they have never been debated. The public has been kept out of science and technology decisions. What we have to understand is the need to involve the public at the beginning of a science and technology revolution, so that we can have as intimate a representation as the scientists and corporations putting the products out. Nothing less than that is going to be acceptable.

PROFESSOR NESSON: Mr. Blobaum, do you agree with that? Is he speaking for you?

MR. BLOBAUM: He is certainly speaking for me in terms of getting involved in the process and raising the appropriate issues. There may be a difference of style in approaching some of these things. But certainly these issues need to be raised in a responsible way, and they are normally raised in coalitions.

MS. CULLITON: What we have just seen is that there are, in general, two issues here. One is the scientific, technical issue, and the other is an issue of social or political point of view.

I would like to ask Mr. Rifkin and Dr. Kornberg right now to show us in a public debate what Mr. Ruder suggested he would like to ask them to discuss in private. Mr. Rifkin has drawn a parallel between the BT apple and the corn experiment. Dr. Kornberg has suggested that the parallel is not valid. One of the things that I as a journalist and a citizen and a potential eater of apples would want to know is whether or not, in this particular case, Mr. Rifkin and Dr. Kornberg could reach any kind of consensus, not as it applies to all of the experiments that might be done, but to this particular hypothetical case.

DR. KORNBERG: I would be very eager to meet with Mr. Rifkin and describe to him each step along the way, and since I do profess to know this field for many, many years and have worked in it very assiduously, I think I can give him the evidence that he needs about how this product was achieved, which I would hope relieve his concerns about its safety.

MR. RIFKIN: This is refreshing to me because in all the years I have been involved, you are only the third person that has come to me. The fact is, only two corporations have ever come into my office; one sending a public relations representative, the other a corporate executive, to sit down with me and explain their product.

Could you do something that these two companies didn't do—provide me with the data that you used to come to your conclusions? I would even be willing to, under advisement, keep that data private and not publish, if those where your conditions. I wouldn't like to do that. I think it should be open to the public, but we cannot just sit there and have you say it is safe without seeing the data and being able to provide it to our scientists. I should say that of the two companies that came to me, one of them, interestingly enough, we thought had a safe product, and we didn't challenge it.

DR. KORNBERG: Let me say that I am not the third person to come to you because I wouldn't come to you. We are doing science at our company. We are publishing our data promptly. It is available to you. I will give you reprints and preprints of our work. It is wide open. We have applied for patents and the work is duly protected for our commercial interest.

Once again, should you call me or appear at my office, I would be delighted to go over all these data and explain the work so that you would be satisfied as to the scientific detail and whatever other concerns you have.

MR. RIFKIN: I suspect you would first have to check with the CEO that you are under a consulting contract with, and I would say they may be very unlikely to want to provide you and me with that data. I can tell you from first hand experience with a couple of companies in this room—let's take bovine growth hormone for instance. We have been trying to get that data for years. It is not publicly available.

DR. KORNBERG: I happen to be a consultant with two companies, and I have full liberty to give you the kind of information that I have offered you.

PROFESSOR NESSON: Mr. Ubell, you kind of agree with Mr. Rifkin's point of view. Is that going to be reflected in the documentary that you put out?

MR. UBELL: I would listen to his story, look at the data that he has amassed, and try to make some judgment. Unfortunately, the documentary does give you a lot of time to do follow-up and work the story, but if you are, like me, generally a daily reporter, you are really stuck, because in the few hours between the end of the press conference and the time you have to go on the air, you have to make a lot of phone calls. That is very difficult to do in journalism, very difficult.

PROFESSOR NESSON: You would be there if this press conference were on Sunday?

MR. UBELL: In New York.

PROFESSOR NESSON: In New York. Just follow it through for me. You call the company. It is in California. It is Sunday. You get no answer. You call all day. You get no answer. You try and track Mr. Salquist down. He is in Pebble Beach on the course. What shows up on the evening report of his press conference?

MR. UBELL: It depends on what Mr. Rifkin is alleging.

PROFESSOR NESSON: Make a good allegation for me, a nice crisp one for the camera, Mr. Rifkin.

MR. RIFKIN: In regard to the BT apple, scientists have raised serious questions about public health and environmental safety that require further analysis before getting a commercial application.

MR. UBELL: That would not satisfy me because it is too general.

MR. RIFKIN: I assume I would have for you petitions, affidavits from scientist that you could then check with.

PROFESSOR NESSON: How about saying something very specific, how about with a little grit for the TV audience.

MR. RIFKIN: I will tell you something very interesting about having a little grit. There is a real problem here that is not created so much by the media but by the corporations themselves. They create a straw man that is a caricature of me and people like me and then they accuse us of being alarmists. They will say things like "Mr. Rifkin says there is going to be a killer tomato in the fields" or "we are going to create a Frankenstein monster" or "we are going to have another Hitler's Germany." Many of the corporate executives in this room that we deal with create that character—they may believe it, but it is not about who I am or what I have said.

PROFESSOR NESSON: So, you are a figment of their imagination.

MR. RIFKIN: In *1984*, "big brother" is always warning against this guy Goldstein, who is the outlaw renegade. He says don't listen to Goldstein. You are never sure whether the Goldstein really exists or "big brother" created him because he needed a whipping boy.

PROFESSOR NESSON: Mr. Ubell, is there any chance that on the broadcast, you are going to get Mr. Rifkin's side of it, and, because you couldn't reach the company, there is going to be something that says the company was unavailable for comment?

MR. UBELL: I would have to get the specific allegation before I did anything.

PROFESSOR NESSON: Just imagine he says BT causes cancer when you eat the apples. I want to know whether you are going to put that on the news without the company's response because you couldn't get it, or whether you are going to hold the story.

MR. UBELL: It depends on the level of news visibility that this product has had. In the Alar situation, we were given very little time to catch up, so to speak, and the result is that the first announcement held the public's attention. In that case, I was able to talk to a lot of other people about it and presented what I thought was a fair and balanced report.

In this case, I would question why he held the press conference on Sunday, and I would say something about that.

PROFESSOR NESSON: Would you say that was a dirty trick?

MR. UBELL: Those tactics have been used before, but I don't think he would, would you, Jeremy—

MR. RIFKIN: No.

PROFESSOR NESSON: Dr. Miller, I have had you sitting there quietly for some time. You are the regulatory authority. Now, all this brouhaha is going on out there. You are reading about it in the paper. You are seeing it on TV. You have an application in. What effect does all of this have on you?

DR. MILLER: In most of these situations, like the guy that follows the horses, I have to sweep up the mess that everybody else is making out there. The ideal situation would be for us to evaluate the data that is being presented to us and come to a reasoned conclusion concerning this material. And this conclusion might either be that there isn't enough data to make a conclusion, or we have questions about safety, or we think it is safe within the meaning of the act; that is, to a reasonable certainty.

The reality of the situation is that the more public the process becomes, the more public the debate becomes, the more vehement the debate becomes, the more *ad hominem* the debate becomes, the more cautious our evaluation will become. We will become more and more sensitized. Regulators are people, after all, and they respond to these outside things as well. The more vehement a discussion is, the greater the interest in Congress, the much more conservative we will become in making our evaluations.

MR. UBELL: Which is what Mr. Rifkin wants.

DR. MILLER: And Congressman Brown wants. That doesn't mean necessarily that we agree with Mr. Rifkin, but the reality is that as human beings, we have a strong, even unconscious urge to protect ourselves against all possible attacks.

PROFESSOR NESSON: Congressman Brown, you have constituents that are just up in arms about the idea that this apple is going to come on the market. They have been bombarding you with campaigns. Will you consider calling Dr. Miller?

CONGRESSMAN BROWN: Under normal circumstances, I would not consider calling Dr. Miller. As a publicity-seeking member of Congress, I would call a committee hearing, invite Mr. Rifkin in, invite the president of the company, and just have a real donnybrook in front of the TV cameras. I would pose as the sensible, rational arbitrator.

DR. MILLER: Precisely, and the group at the end that will take the greatest blow will be the regulators. On the one hand, they will be accused of taking an inordinate amount of time and keeping off a vital technology, and on the other hand, by Mr. Rifkin, Mr. Blobaum and others, of not doing a credible review, not listening to the right people, not having the public interest at heart. In other words, our decision is being made too quickly. This is part of the regulatory game, but, nevertheless, in the end, no matter how long it takes, our review will be as thorough as we believe it has to be in order to have us reach a comfort zone of safety.

The hardest problem in this business will be to define the questions we have to ask about this apple. This is a very difficult issue, and while Roger Salquist and Arthur Kornberg and Roger Blobaum will be confident about this thing, the questions we ask are somewhat different than the questions that they ask.

For one thing, we have to ask ourselves who we are protecting. Are we protecting the average, traditional, 25- to 50-year-old healthy male individual, or are we protecting young children, or are we protecting elderly people? Are we protecting people who eat a reasonable amount of apples, or are we protecting people who eat a hundred pounds in the course of a week? The answer is a question of judgment and our job is to protect the average citizen. Our job is to protect the fool, but I can't protect the double damn fool. That is where the difference lies. I can't protect the person that is going to eat a hundred pounds of apples a day.

PROFESSOR NESSON: But the old people, the young people, the sick people, that whole spectrum.

DR. MILLER: That I have to worry about. By the way, there is nothing in the law that says this. This is a matter of personal philosophy and that is part of the problem. The law in many cases is highly precise.

PROFESSOR NESSON: Mr. Blobaum wants to say something to you here.

MR. BLOBAUM: Yes. I would like to know whether you would look at this in a more thorough way, because it is a precedent-setting kind of case. Would you consider whether or not this is a food additive, for example, and other more creative ways of looking at this, because this is the first time?

DR. MILLER: I need to be comfortable in that I know something about what is going on. I need to feel comfortable that I am asking the right questions, and I need to be comfortable that I can make a decision in which, if I am wrong, the consequences will not be very significant.

PROFESSOR NESSON: Let me just pose it to you in terms of our hypothetical. You have a significant part of the public up in arms about this apple, and what they fear specifically is that this BT protein, this genetically engineered apple with the "mutant" protein in it, is somehow going to get loose in the environment. It is going to spread like Johnny Appleseed. The birds are going to eat these apples. People are going to eat them. We are eating genetically engineered, or mutant stuff. Now, you look at all the data and you get convinced that that is not valid, but you have the American public out there, and a *USA Today* poll indicates that 67% want you to keep this apple away from us, 22% are for it, and the rest are undecided. How does that affect your judgment?

DR. MILLER: You asked if I was convinced it was safe, all right? I am comfortable that it is safe. I can't ever be sure that it is absolutely safe. That is why the hypothetical here that talks about no risk is foolishness. There is no such thing.

My recommendation would be to approve, in spite of what the polls are telling me. The reason for that is pretty straightforward: I could never defend a decision in court not to approve this apple because the public believed it was unsafe. If this company sued the agency and said, we think this is safe and you are inappropriately not permitting the use of this apple, then I would have to go into court and defend my decision on the science, not on the basis of polls.

In 1958, Congress, in its infinite wisdom, specifically informed the agency that it was not to make decisions on the basis of things like need and popularity, only on the question of safety. Safety is determined to a reasonable certainty. Even Congress could recognize absolute safety was not possible.

PROFESSOR NESSON: Mr. Rifkin, he is about to approve this thing. Can you stop him?

MR. RIFKIN: If we felt that the review process wasn't adequate, we have several courses of action. One is litigation in Federal Court. Two is to allow them to go ahead with the commercial application and mount an effective public education campaign and boycott. Three—and this third alternative is quite interesting because it allows the Food and Drug Administration Commissioner the ability to deal with both sides—there is a provision in the FDA regulations that says that if a significant proportion of the public wants a particular product to be labeled, the FDA has to take that under advisement, and perhaps put labeling on it. I would suggest in this instance that if you get a product like bovine growth hormone and it is approved, those who are upset about it may want to petition the FDA, which is a legitimate right, although many of you would think that was inappropriate, and ask that it be labeled as a genetically engineered product.

Then those members of the public who would like to take the product will know that it is genetically engineered, but they will accept it because the FDA said it was safe, and those citizens who feel that they cannot trust the FDA decision will look at the product. It will say genetically engineered on it and they will go for an alternative.

That is the best possible scenario I could imagine because then the marketplace will decide.

MR. SALQUIST: That is our best scenario, too. We are going to label this thing. That is why people are going to buy it. This is superior to an apple that has been treated with pesticides, albeit the residues may be small and totally within tolerances.

MR. RIFKIN: Do we have you on record this morning saying that you are going to label this product? How about your tomato at Calgene, are you going to label that? Let's get into reality.

MR. SALQUIST: We are not talking about that. We are talking about a hypothetical—

MR. RIFKIN: Well, wait a minute. You should be sincere in your beliefs. If you believe it hypothetically, will you label the tomato as genetically engineered? Yes or no? Let's be at least honest.

MR. SALQUIST: Will we label the BT apple? You are damn right we will.

MR. RIFKIN: Well, then label your tomato.

PANELIST: You will have to label everything because every fruit and vegetable is genetically engineered.

MR. RIFKIN: Well, we have a problem because you have products like L-tryptophan that may have been caused by a recombinant DNA process, and if it turns out to be the case, you have a lot of food additives out there in other processes that are being designed with recombinant DNA that could be damaging to public health.

PROFESSOR NESSON: Here is the way things have gone here. This process didn't go nearly as fast as you hoped it was going to go. The public pressure on the FDA has resulted in their being extraordinarily cautious with this product, and, in fact, it looks fully as if it is going to be years from now before it gets approved.

Your bills for PR are substantial. Your legal bills are mounting. You have a lot of other products in this company. Your company is getting knocked on all of its products by the bad publicity around this BT apple. There is frankly not that much money in it, even if you get it out. Would you consider just bagging it and going on to something else?

MR. SALQUIST: Well, absolutely. If the rate of return at some point goes negative, and you are throwing good money after bad, hell, yes; you abandon it.

PROFESSOR NESSON: Even though you believe in it?

MR. SALQUIST: Look, I am in the business to make money. I am not on a religious crusade to have this technology out on the market.

PROFESSOR NESSON: All right. Now I would like to finish this panel in the following way. Congressman Brown, I want you to hold your hearing. All right? Not on the BT apple and not just for your publicity-grabbing instincts, although I don't want to deny those to you either. I would like you to hold a hearing right here for this group on the subject of science safety, risk, and the public perception. All right?

Let's say that the premise for your hearing is that this business of putting out new products, technologically innovative products, is getting distorted by the intensity of advocacy on both sides, and that there has got to be some better way to bring the public to an involvement that is more informed, bringing industry, advocacy groups, and the public interest into better focus so that Dr. Miller and the other agencies can do their jobs better.

I would like you to invite some of our other distinguished participants to be witnesses before your committee. Mr. Dunlop, you are the first witness before this committee. You are an expert in regulation. We describe here a regulatory process that is at least vulnerable. Do you have any suggestions, suggestions for Congressman Brown as to how we might proceed in a better fashion?

PROFESSOR DUNLOP: I would suggest that Congressman Brown explore the possibility of a wider introduction, in the food area and the FDA, of negotiated rulemaking, as is happening in OSHA and the EPA. There, all the facts are gathered and all the parties sit around and try to work it out, in no way, however, impugning the responsibility of the administrator. I don't see why the food area can't follow what we do in other areas.

CONGRESSMAN BROWN: I will.

PROFESSOR NESSON: Your next witness is Mr. Shapiro, CEO of Monsanto Agriculture.

CONGRESSMAN BROWN: You have heard the suggestion from Dr. Dunlop. Do you think this is a feasible way to get started in the regulatory process?

MR. SHAPIRO: I think it is worth exploration. I wouldn't go into it with perhaps any great optimism, but I think it would be a good first step.

CONGRESSMAN BROWN: You did hear Mr. Ruder say that he would have preferred to begin this process on behalf of his company by doing something similar to that, by bringing in the various parties at interest here for a discussion, didn't you?

MR. SHAPIRO: I think that is clearly an enlightened and highly practical approach. It is one that most corporations have not historically felt comfortable with. The historic notions of product development and product approval have tended to focus on a closed process, one in which competitive information is guarded carefully, one in which sharing information is seen as a possible threat, both in the public arena and in the competitive marketplace. But I think Mr. Ruder speaks with wisdom.

CONGRESSMAN BROWN: And the bulk of the corporate CEOs are not quite as enlightened as they should be?

MR. SHAPIRO: I think corporate CEOs are learning a lot pretty quickly these days.

CONGRESSMAN BROWN: Do you see any way in which you might involve some other elements of the public in this discussion? For example, there are going to be a substantial number of members of the public who feel that the BT apple or Adam's Apple, as it has been renamed, might be inherently evil because it was the apple that Eve gave to Adam, that led to the fall of man.

MR. SHAPIRO: Without getting into theological debate, one of the reasons that I am not entirely optimistic that this process is going to produce reasoned consequences—or even mutually acceptable if unreasonable consequences—is that issues of fact generally can be discussed until we have reasonable agreement as to what we know and what we don't know. Issues of political perspective and value are much harder to reconcile through the process of public discussion.

I think it is clear that some of the differences in this debate are not issues of science. They are not issues of risk. They are not issues of fact. They are issues of profound world view. And I am not thoroughly optimistic that the regulatory process is a good place to resolve those. That is why I favor Dr. Miller's approach, which says that under the current statutes he is restricted in his decision-making to dealing with issues of public health. He is not there to decide need. He is not there to decide issues of broad, long-term planetary implications, other than issues of health, or in the case of the EPA, issues of the environment and so on. As long as we can keep the regulatory and governmental role focused on those, and allow the media role and the role of public debate to deal with the broader issues, I think that is a workable framework in a democracy.

PROFESSOR NESSON: Dr. Caroline Jackson, you are involved in issues of regulation in the Common Market. Would you give the Congressman the benefit of your perspective on this American scene, where public opinion seems to play such a prominent role in regulatory activity from the perspective of the EC?

DR. JACKSON: From the perspective of the European Community, two things strike me about this debate. One is the intensity of the debate outside Congress and the enormous pressures that can be raised by extra-Congressional groups and people like Mr. Rifkin.

But the other point is the way in which BT has been developed, apparently, without any reference to international considerations and considerations of international trade. As a member of the European Parliament, I will, of course, be raising the issue of BT in the Parliament. I will be calling for a special committee of inquiry, possibly in parallel with the one that is currently running in Congress. And we will be asking the Agriculture Commissioner to consider a ban on American apples if, in fact, BT is developed without any reference at all to international standard setting or to European standards.

CONGRESSMAN BROWN: But Dr. Jackson, you certainly can't feel that there are any problems of health and safety with the BT apple.

DR. JACKSON: Well, I know that you have great respect for the findings of the FDA, but, of course, the findings of the FDA are only relevant immediately in the American regulatory context. We in Europe don't want to start from square one when we consider BT. We will have the FDA findings and recommendations on file, but the EC's Scientific Committee for Food, a number of other regulatory bodies with the European Community, and the European Parliament will want to look at all the evidence. I think what Congress needs to take into account here, considering the likely future of BT in the European context, is the precedent of BST, and the successful pressure, originating in the European Parliament, that has resulted in the continued moratorium on the use of BST. And remember also the successful campaign launched by the European Parliament against the use of growth hormones, which has had an enormous impact on U.S. trade with the EC in terms of beef.

CONGRESSMAN BROWN: Dr. Jackson, how can you possibly justify the action of the European Community on BST when it had been declared safe after exhaustive field hearings in the United States?

DR. JACKSON: Well, we are always very interested in the exhaustive field hearings in the United States, but we also wish to carry out our own tests, and we felt that there was not sufficient evidence considered by European sources that would enable us to be 100% sure as politicians that BST was absolutely safe. We couldn't recommend that to our constituents and, of course, we were well aware that there were quite independent consumer groups operating in a number of European Community member states who did not want to have BST-treated milk released onto the open market. That was something which we had to take account of, as we would have to take account of the growing movement within Europe that is going under the slogan, "Eat Natural, Eat European."

PROFESSOR NESSON: Ms. Jackson, I hear you rather explicitly saying that from your point of view regulation has considerable political perspective—an overt component of the process, in fact. In the United States, Dr. Miller says "Safety is my only guide. I feel these pressures." You actually regard the process as one of political regulation?

DR. JACKSON: Yes, that is true, and I would also point out that in the European context, Mr. Rifkin would be a member of a parliament or a member of the European Parliament. We have our "Greens" and our environmentalists inside our political process, rather than agitating from outside.

PROFESSOR NESSON: The next witness before our committee, Dr. Smith from Nabisco. Dr. Smith, do you have anything to offer the committee by way of wisdom on this subject?

DR. SMITH: Well, clearly, my concern is that we now have 11 groups of scientists. I think we are going to run out of scientists. Is there any way that we could have some kind of a superscientist oversight group? We used to depend heavily on the National Academy of Sciences and some of these agencies that had a lot of credibility and could view some of these questions more dispassionately. I would favor moving back into that arena because I think this business of our choosing our own scientists, the ones that agree with us politically and otherwise, is very risky.

CONGRESSMAN BROWN: Dr. Smith, your suggestion is attractive on the surface, but the National Academy of Sciences has demonstrated over many years that they can't even agree on how much fat in beef is healthy for the American people. You are surely not expecting them to solve a difficult problem like the BT apple.

DR. SMITH: There are a lot more hands going up. Maybe you have to call on some other witnesses.

DR. KORNBERG: Mr. Congressman, I think you have just inverted the difficulties. In the case of the BT apple, the data are clear and explicit. We don't have any problem with regard to evaluating the data and I think I can convince you of that. With regard to diet and cancer, we have frankly told you on many occasions that we don't have the information to tell you what to eat or what not to eat.

CONGRESSMAN BROWN: Mr. Rifkin, I will recognize you briefly.

MR. RIFKIN: I personally have been in front of congressional committees for years on the question of predictive ecology for genetic engineering. There is no risk assessment science. There is not an adequate database to judge the potential risk of releasing any of these genetically engineered products in the environment, and for years various groups, like the National Academy of Sciences, the National Science Foundation, and the EPA have pledged that they would begin developing a sophisticated risk assessment science. It doesn't exist. Therefore, in conclusion, how can we possibly regulate a product line when we know there is no science from which to do it? It does not exist.

PROFESSOR NESSON: Dr. Smith, do you actually believe that we could assemble a panel of scientists who would be both unbiased and perceived to be unbiased?

DR. SMITH: No, I would be looking for a balanced scientific board that would look at the issue from several biases, if you will, but come to a consensus, so the public would trust their findings.

PROFESSOR NESSON: One of Mr. Rifkin's scientists and one of Mr. Salquist's scientists.

MR. SMITH: One of the 11 groups of scientists—I mean, they would all have to be represented.

DR. MILLER: Scientific advisory committees, *ad hoc* scientific advisory committees, have been a tool in the hands of the regulatory agencies for some time, to deal with difficult issues. As a regulator, I would be much opposed to the idea of submitting all decisions of the agencies, at least my agency, to a scientific advisory group. If that were the case, a process that is already almost biblical in its length would become even more so.

MR. BEATTY: I would like to say that I can bring it back to a commercial full circle. We are, in Canada, the largest purchasers of apples for inclusion in apple pies. We dominate that market, and we would not allow the farmers from whom we purchase our apples to begin experimenting with inputs to our food products that were regarded as hazardous by any section of our consuming public at all and, indeed, would expect that if we didn't exercise that discipline on our farmers, our customers, the supermarket chains, would do it for us. If there were, in the end, a tremendous public uproar about Mr. Salquist's apples, I can assure you that as CEO of a food manufacturing corporation, we would not buy any such apples from the farmers involved.

CONGRESSMAN BROWN: If that public uproar were based upon a nonscientific, nonrational reason, such as—and this is very possible—religious objections, the fear that the apple might be evil, would that still lead you to take the same course of not using the apples?

MR. BEATTY: I am in the business of producing apple pies that people will buy, and I will make the safest decision for my consumer and for myself as a manufacturer that I can.

CONGRESSMAN BROWN: Would you be willing to participate in a process that would lead to a change in the public perception, since it was essentially nonscientific?

MR. BEATTY: No, because I think we have seen that there are armies of scientists arrayed on every side, and where there is uncertainty, I expect that men and women of good will are arguing from bases of profound belief and understanding. It is not my job to adjudicate among 17 competing scientific groups.

PROFESSOR NESSON: Assuming that you think the science is on the side of the BT apple, and the public perception is being manipulated by cynical public interest groups, out to raise money for their own coffers—you completely buy Mr. Salquist's view—the question is: would you participate in an effort to educate the public?

MR. BEATTY: I doubt it. It is simply not my business.

MR. SALQUIST: If we sold more apple pies for you, you probably would. I mean, the only reason I could see that you would have a positive answer to that is if for some reason this was construed that there were going to be a lot more apple pies sold.

MR. BEATTY: A marketing advantage derived from the miracle new product.

MR. SALQUIST: Sure. That is the only reason you would want to do it, and that is the only rationale that I would come to you and be eliciting your support for.

MR. SHAPIRO: Let me point out one consequence of that position, a perfectly understandable commercial position, which says if my customers want it, I will produce it, and if they don't want it, I won't. We have moved from a regime of governmental regulation about what is permissible, to a regime of public agitation on the subject of what is permitted in the food supply being the ultimate criterion for determining what is available to the public. The whole point in having a food regulation system was to resolve scientific disputes about safety matters that were legitimately disputable. At that point people could put those approved products on the marketplace and then consumers could decide for themselves whether they wanted to buy them or not.

But if you are going to make that decision based on the possibility of controversy, then the people who do want to buy them don't get a chance to buy them; the people who do want to use it in growing don't get a chance to do that. You have allowed another regime to substitute for the one that Congress and the legislatures of other countries have put in place.

MR. BLOBAUM: I would like to support Mr. Beatty's position. I think that the consumer is the ultimate decision-maker in this whole process. If the consumer is not satisfied that the process has been fair and fully explored and carried forward to a satisfactory conclusion, all the consumer organizations that I am familiar with maintain that there needs to be choice in the marketplace. If people do not want a BT apple, they should be able to buy a non-BT apple. What will ultimately happen in this kind of case is that there will be non-BT apples on the market. That is a foregone conclusion. The consumer is the ultimate decision-maker, despite all the science.

PROFESSOR NESSON: You are suggesting a world in which I go out to buy apples, and some of them have little BT labels and some of them have no BT labels. And when I go out to buy apple pies, the apple pie has a no BT label on it and some others have a BT label. In the world of apple pies, what would conceivably lead me to buy the BT-labeled pie over the no BT-labeled pie? As a consumer, why take a chance? There is this controversy out there. I might as well be safe. If I want it to be natural, stay clean, I will go with the "no," rather than the "yes."

MR. BLOBAUM: I would say the consumers would go with the no BT apple. The other side has to make their case as to why this is a better apple, and this is one of the flaws with this product. It is not a better apple.

MR. BEATTY: We have introduced in the Canadian marketplace an organic bread. If there is a sufficient demand for the product, if there is a sufficient number of people out there who wish to eat products solely grown on lots that haven't been pesticided or insecticided for a certain number of years, we will seek to serve that market. And if for some reason somebody should come forward with a BT apple that has some attribute that is particularly desirable in the consumption of a pie, I would anticipate that we would produce an apple pie that would be so labeled for that segment of the population.

But I don't see that we have a moral imperative to enter into the scientific debate regarding whether or not this is a sound process, and the fact that there is debate, discussion, confusion, altercation, and a lot of harsh feelings indicates to me that it is an area that as a processor I am very wise to stay well clear of.

PROFESSOR NESSON: Mr. Goldberg, the problem seems to have focused at this point on the fact that all of these decisions have heavy political consumer components to them and that at times it will be the perception of those who might be considered elite in the science business that the political consumer view is an uneducated view. Do you have any approach to offer the Congressman as to how we might better deal with that problem?

PROFESSOR GOLDBERG: I agree with the witnesses who have indicated that the food system, for better or for worse, has environmental, nutritional, and consumer components to it, and as an academic trying to train young men and women for that food system, I would like to see this generation and the new generation find a forum for involving all these components in the food system rather than having them remain outside. I don't think we have to wait for a Green Party, like the European Community, to get them into the system. I think that there are informal forums in other industries that have worked to involve these groups, and it doesn't have to be a confrontational dialogue. It can be a constructive one.

I wished you had called Mr. Dunlop, your first witness, last. As far as I am concerned, this gentleman has been able to create such forums in other industries and I would like to suggest that you hire him as a consultant to create one here.

CONGRESSMAN BROWN: Mr. Goldberg, you are endorsing a proposal made by Mr. Dunlop, which Mr. Ruder referred to earlier as being a process that he would feel comfortable with. Is that right?

MR. GOLDBERG: I am endorsing two proposals that were made. One was a governmental negotiating type of proposal that led to negotiated settlements in a governmental framework. I am talking about a nongovernmental framework as well, outside of that system, that would be a continual dialogue, not on an issue by issue basis, but one that would bring groups together on a permanent basis to discuss where the industry is going, how it is going to get there, and how we can do a better job to use this creative science that is going to do so many wonderful things for us, rather than worry about how it is going to destroy us.

CONGRESSMAN BROWN: Do you see this process as creating a more favorable public climate for the acceptance of the results of the regulatory process?

MR. GOLDBERG: I think that the two go hand in hand.

MR. RUDER: I think, Roger, if we start off with the premise that we are dealing with reasonable, well-intentioned people in the first place, and if we are willing to talk to each other before things happen rather than after things happen, and avoid our boxing each other into the classic cat and mouse kinds of positions, we can make a lot of progress.

I suggest that we can institutionalize that process and, Roger, it just seems to me that you ought to give some thought to getting together with some of our fellow panelists to talk about institutionalizing an approach—a methodology—that will make sure that our talking past each other doesn't happen in real life.

PROFESSOR NESSON: All right. Now, let me offer some thanks here. I want to thank Jeremy Rifkin for coming. I think that this is a tough house for you to walk into and I want to thank all of the members of the panel, but I want to note specifically that I think that you helped make this panel go by being here and offer you thanks for that.

PART VI

THE GLOBAL FOOD SYSTEM: INDUSTRIAL PERSPECTIVES

CHAPTER 19

Business-Government Relations

John T. Dunlop

Our discussions started with the continuing and the future role of new technology, which, of course, keeps changing the problems we confront. And we have learned that this new technology is operative throughout the food chain, in agriculture, in processing, in distribution, in packaging, in preparations throughout. It is not simply at the early agricultural stage. We also saw that the food chain is global and has, in many respects, been global for a long time. The Common Market, Poland, and Mexico illustrate the global qualities very well. There are numerous international interactions in these areas.

We moved on to the conflicts between or among scientific standards and their comportment and the needs of society; the issues of politics, regulations, and various interests that have been dramatized earlier. While you might say we have highlighted those differences, I, for one, am often impressed with the degree of commonality. We must always recognize the important commonalities within those ideas.

There are two ideas that I want to make explicit: The first is that in the governmental regulation process, we have come a long way in the last 15 years, discovering new ways of regulating. In 1975, we began to introduce in this country something called negotiated rule-making, which is quite different from the rules of the Administrative Procedures Act. The Administrative Procedures Act requires administrators and staff to draft regulations, seek public comments upon them, which are often extreme, and then look at whatever they want to look at and come up with a new set.

Few have written as many regulations or signed as many as I have over 50 years—virtually all under the Administrative Procedures Act. We began in the mid-1970s to introduce rules through negotiated rule-making, bringing in all the facts, and bringing in the various interested groups—environmentalists, consumer groups, labor, management—to work these problems out. Not that the regulator would ever yield his authority to issue regulations; he would put what he felt was a consensus, if he wished, into the *Federal Register*, so anybody in Missoula, Montana, or elsewhere could make comments. The final decisions would, of course, be by the regulator, subject to court review. That process has assured much, much less litigation. We have introduced this in places like OSHA and EPA, not to mention the cases of negotiated rule-making I developed myself in the coke oven standard and mass transportation regulation.

In 1990 the Congress and the President enacted a new law authorizing the use of negotiated rule-making (P.L. 101-648). It is my hope that industry—in its environmental, health, and food areas—will learn from this newer form of regulation by negotiation. It is possible under our law and under our processes to apply faster, less litigious, and more responsive methods of regulation when appropriate.

The second idea I wish to suggest is a very different one. It has nothing to do with the government. A number of industries I am associated with have developed the concept of a channel, if you like; that is the word that is popular in business schools these days. We had one such session recently in the shirt industry. People came together: the CEOs of the yarn companies, the CEOs of the textile companies that make the cloth, the CEOs of the shirt companies, the CEOs of the retailers, and the union leaders that represent the workers in much of the channel. This group set the stage for a private discussion of a range of problems through the whole channel. The issues of communication, computerized data systems, and inventory replacement through that channel are matters that are discussed in many such segments.

I want to distinguish the first view, where government has an important inevitable role as regulator, from the second, the sessions to which the government is invited on occasion, but where it is not the mover and shaker of the discourse.

We now turn to men of great experience and practice, leaders of different segments of the food industry. What follows is an examination of some major problems facing this industry and the world food system.

CHAPTER 20

Food Packaging and Government Regulation

Harry A. Teasley

I shall comment on another link in the food chain: packaging. A lot of what is being said about food safety, and about the political pressures, public relations issues, and the issues of fact and science that relate to pesticides and biotechnology, also relate to packaging, but from a different point of view. Instead of deriving from the point of view of human safety, these issues relate to the point of view of environment.

There is a real need for harmonization and for food regulations to facilitate world trade. Right now in the United States there are literally thousands of proposals and bills and enacted legislation on the books that would vulcanize the American food industry by making packaging state-specific. I would like to give you a flavor of the kind of regulations that are being proposed.

One proposal is that plastic packaging should be banned because it is made from a nonrenewable resource. The implication is that glass is better. However, if one does a systems analysis and an energy balance from extracting materials through disposal, one finds that plastic packaging actually has a lower net hydrocarbon usage than glass (including the hydrocarbons in the plastic itself). Another proposal is that multilayer packaging should be banned because it may be viewed as nonrecyclable, since it is composed of several materials. But, again, multilayered packaging has been one of the great ways to reduce the material and energy content in packaging by tailoring a package to get a characteristic that is needed, like oxygen protection, flavor protection, or vitamin C protection, in a specific material and combining them. Tetra Pak faces exactly that issue in Maine.

Another contention is that all packages should be made of materials having a minimum recycled content. Some materials are easier and more economical to recycle than others. Some applications have different technical needs than others and, yet, there is a desire to put out a "one-suit-fits-all" kind of proposal. It is also proposed that packaging should be made and sold with mandatory refunds, and that is really controlling the terms of sale or the price of products in the marketplace.

And lastly, some say that the government should specify materials and packages for each product application or end-use application. There are literally millions of them, and that would require a kind of information system omnipotence at a central level that I don't think is possible. Yet, all of these regulations are clearly out there in different forms, and there are many, many more.

I would like to take an example and walk you through it. It is an example from the orange juice industry. I was at a seminar, when somebody attacked me because they said oranges are the best example of no packaging. Mr. Teasley, you shouldn't have packaged orange juice. Oranges do require packaging. They are shipped to a retailer in a rather heavy telescoping case. You ship the skins and you ship all of the water, and the numbers are quite interesting. At the retail level, there is corrugated waste produced with oranges that is substantially more than that for concentrated orange juice, by a factor of about 8–10, and at the consumer level, there are wet skins that produce more waste by a factor of about 60. So, just on that basis, there is an argument for packaging. But another interesting argument is that at the industrial level, we squeeze oranges more efficiently. It takes the consumer about 25% more oranges to get the same juice that we get out of an FMC squeezer, which consists of knives that interlock like a punch press. They really hit an orange; we get all the juice. Not only that, we reclaim the oil and sell it for flavorings, we get delaminine, and we convert the skins to animal feed. So, we end up using all the orange and producing no waste.

We are very efficient, then, with agricultural resources, pesticides, herbicides, agricultural capital, and agricultural labor vis a vis free oranges. We also can ship six or seven times as many gallons per truck as fresh oranges. I would argue that the marketplace is not stupid. Over time, competitors in the marketplace find very efficient ways of doing things, and prices are more than a mechanism for transferring goods and services. They are an incredible means of transferring information about the resources that may be contained in a product or service.

I am not for banning fresh oranges. I think you should have the option of buying fresh oranges, but the price of those fresh oranges for use as orange juice is going to be higher than the price of the products that I can deliver to you.

Frozen concentrated orange was invented in 1946, and the man who invented it had a hard time selling the juice. He went door to door in his town where he lived near Boston, and over time he built an incredible industry. Should that process have been allowed in 1946? The product wasn't that good. It had a burned character, but it made orange juice available year around. Prior to that time, oranges were a Christmas stuffer for a lot of us, even in Georgia next to the state of the Florida, because they were a seasonal product and there was no way to get them.

Since that time, we have developed high volatile essence recovery; low oil extraction; freeze concentration; high-temperature, low-time evaporation; and dramatically improved the quality of the product. And if processes don't have safety problems, such as food irradiation, I don't think that we should have government saying maybe we still shouldn't have the process. I think the marketplace will decide, in due course, what processes and products should exist and what products and processes should not exist.

CHAPTER 21

Sourcing Fruits and Vegetables in a Global Food System

John A. Woodhouse

My organization, Sysco, is a pure marketing and distribution company. It is fundamental to our principles that Sysco not be involved in manufacturing or processing. We leave that to the roughly 3500 great food processors of America and around the world, and we rely on them to produce approximately 175,000 individual items that we have in our company-wide inventory, all of which are going to 230,000 customers—essentially what we describe as the away-from-home eating market in North America. As a distributor, we are that middle man you hear about, consigned to the economic trash can as useless by some dubious economists, like Karl Marx and some of the early classical economists.

It is from this perspective that I comment on the world food system. One of the most noticeable changes I have observed over the last 20 years is the impact of foreign food supplies on the whole United States food system. Two decades ago, a company like ours marketed only a handful of products that were produced abroad, and they fell into two categories. There were the dirty and undesirable labor-intensive products, like raising and cultivating mushrooms, or harvesting pineapples. Nobody really likes to do it, and it gravitates toward where you can get cheap labor. The other were products that had to rely on unique climatic and/or soil conditions. Today, it is remarkable that a company like ours is securing hundreds of products from nearly 50 countries around the world.

Much of this reflects the introduction of United States production, processing, technology, and know-how around the world. Essentially, we have learned to export our agricultural know-how. For example, the seedless red bunch grape was initially developed by people in California. It was wonderful for use in hospitals and school systems, where it provides nutrition without the danger of the seeds. Today, a great deal of those seedless red grapes that our school children are eating are grown in Chile because a California product could be produced six and seven months a year and the alternate season could come out of South America. Incidentally, the same people who developed it and some major landowners in California are arranging for production in Chile.

Another example: in spite of the fact that California produces over six and a half million tons of tomatoes a year, our company is now importing tomato paste from Hungary. It is of extraordinary quality and superior viscosity, and is quite distinctive. A company over there dedicated itself to excellence, and we are delighted to market its product in the United States.

Many of those quick frozen vegetables that used to be an exclusive of the Salinas-Watsonville area in California are now sourced abroad. Much of the quick-frozen broccoli and some of the frozen peas we market are now grown and packed in Mexico and Central America, in some cases to benefit from the economics available. What has happened in a couple of cases, interestingly, is that California packing plants were disassembled, put on barges, shipped down to the West Pacific Coast ports of Mexico and Central America, trucked into the interior, and set up. Obviously, you end up with a product grown in a satisfactory growing region. Not as ideal a region as California, but with the low wage rates and the labor you can utilize, you can have just as superior a product.

Two major matters concern us as we go abroad. First, we have to do business on a long-term basis. Once we start utilizing any foreign or domestic source, we intend to maintain a mutually supportive relationship for quite some time. It is self-interest, for we are dealing ultimately with an agricultural product that is going to have variable qualities in certain years, and, likewise, variable production levels. Long-term relationships are the only way to overcome the variability of Mother Nature.

Second, overall quality is quite a different concern outside of North America. We utilize some 110 inspectors, most of them really food technologists, in the United States and abroad to assure that the products we market meet our standards (which exceed or match governmental requirements). Surprise visits to plants and fields, and ongoing education of people outside the United States, are vital to make this effort work. Worldwide, only one out of every three plants that we investigate can meet our needs. Some of the reason for this is style. We are used to having specific counts of products; abroad, the acceptable range of what is put into a can is much wider. But other reasons are purely sanitary in nature. In Asia, we can only approve about one out of every five plants; all of us need to realize that United States sanitation standards are far from having worldwide acceptance.

Let me conclude by relating one incident that shows the complexity and interrelationships in this worldwide food industry. At one time, all canned apricots, an item that is intensely used by the hospitals, school systems, and nursing homes of our country, were produced in the central valleys of California. Over a series of years, due to bad crop years, some changing use of land, increasing costs of farm labor, and then, of course, water and environmental matters, apricot production declined significantly. Prices started to rise, and we weren't able to get all the product that we needed for our customers. So, Sysco turned abroad. We found one of the best alternatives available in the valleys of Spain. We identified quality packing plants. They inspected well, and, over the years, these producers could pack the style that we needed for our customers. But to secure economical freight rates, which means multi-containers from one source, we needed to procure more than just apricots. So, we added tomatoes from Spain. This gave us the available quantity to get appropriate freight rates.

As all of us in this room are aware, the European Community made a decision a couple of years ago to ban the importation of U.S. meat using a hormone. After futile negotiation to remove the ban, it wasn't surprising that our government chose to strike back. So, it raised the rate on imported tomatoes to almost a 50% duty. Obviously, this makes tomatoes uneconomical to import. Thus, we no longer can get the critical mass of product to justify bringing in Spanish apricots. Today, our business in Spain has dropped significantly. Investments that people in that country made in good faith are underutilized. Spanish farmers and the Spanish packers are hurting, and an organization like ours, committed to supporting our customers' menus with the products they need, is sourcing apricots and associated products to the Marrakech region of Morocco. I doubt if anybody ever understood, when the decisions were made, how they were going to affect the worldwide food system. From my position, this illustrates the vast complexity and increasing interrelation of the world food sources.

Innovative individuals and firms who understand this complexity can utilize it and adapt to it. They will adapt to it to meet consumer needs, earn a return in the process, and get the products that are required to market in the most efficient manner possible.

Behind all of our discussion, you will recognize that we add the cost of regulation on to the product. The people who really pay for regulation—and they pay in tenths of cents, but those add up—are the ultimate consumers of the product, whether they are paying in the grocery store or in the restaurants and college food systems.

My guess is that what I have described is going to accelerate. Know-how does slip across borders, and know-how is what is involved in most of this. You can overcome difficulties for a lot of climatic problems. You can compensate in many other ways, and I suspect the world food industry within 10 years will be increasingly interrelated, crossing the borders of various countries.

CHAPTER 22

Government Regulation of Health Messages

Arnold E. Langbo

In 1984 Kellogg, with the support of the National Cancer Institute, introduced health messages about the value of a fiber-rich diet in advertising. There has been a lot of controversy in terms of whether that was a message or a health claim. We saw it very much as a health message, although I was not directly involved in it at the time. Fiber awareness, prior to the 1984 campaign on All Bran, was lower than 30% with adults in the United States, and just four years after the start of that campaign, it had risen to more than 80%. Cereal consumers were eating a lot more high-fiber cereals. They increasingly continue to do so—not only cereal, but high-fiber breads, and other whole grain foods that are rich in dietary fiber. Additionally, in 1989 a Federal Trade Commission study showed that new cereals that had been introduced since 1984 were higher in fiber, on average, than those introduced prior to 1984.

There is considerable controversy surrounding this situation, even with respect to how Kellogg handled it. If we were doing this situation again today, we might do it somewhat differently. But at the time, science was overwhelmingly supporting a higher fiber diet, and the benefits, we thought, overwhelmingly overshadowed any risks. The National Cancer Institute, at least at that time, agreed with us, and after all, this All Bran product had been consumed for some 60 years all around the world. Furthermore this increase in fiber awareness did not come from the Food and Drug Administration, the Federal Trade Commission, or any consumer group. None of those organizations told Americans about fiber. It was brand advertising, and we continue to believe that consumers were well-served in the process. I guess you could argue about what happens if children were to eat exorbitant quantities of All Bran, but I think those of you who know this product won't worry that kids are going to eat too much.

I would like to talk briefly about another situation that has us perplexed. There is a grain called psyllium that is grown only in India. It has been consumed by Indians for hundreds of years. For a number of years, psyllium has been available as an over-the-counter drug in America, in the form of a product that you mix with water and drink for improved regularity.

Through research, we discovered some time ago that this grain had eight times the soluble fiber of oat bran, and in clinical testing, a cereal with approximately 15% psyllium had a significant serum cholesterol-lowering effect, in addition to its laxation benefits. Psyllium at the time was not GRAS (generally regarded as safe), only because it hadn't been used in food products to any great extent. The Food and Drug Administration had never really faced up to the issue of whether or not psyllium should be GRAS.

The FDA regulatory system allows us as a company to make a self-determination of GRAS. That is what we did, and then we filed a petition with the Food and Drug Administration for their affirmation of GRAS. If the Food and Drug Administration believed there might be health risks, we would be asked to cease the sale of the product during this process.

We introduced the product. We filed a petition for GRAS affirmation, together with literally a truckload of data to the FDA, and that was over two years ago. We are still waiting today for a decision either for or against the use of psyllium in a food product. We have not been able to advertise the product to any extent because we are not sure that it is going to be around for a long time. But with all the up-front research and development work, we have enormous sums of money tied up in this process.

While the FDA has been pondering the safety issue, the state of Texas came forward and accused us of selling a pharmacological drug product and asked that we discontinue sale of the product in Texas. Now, the FDA has been preempted by the Texas, but we have continued to sell the product because we believe it is the jurisdiction of the FDA to deal with this issue at the federal level. Just the other day, Canada approved psyllium for use in a food product at higher levels than we are using it currently in the United States, and they asked us to put a claim on the product about its laxation qualities, just so consumers will know about it. They also asked for an allergy warning, which we already have, and maybe even a child warning. We can do that. Incidentally, the same firm that originally introduced psyllium as a laxative drink product now has introduced a wafer biscuit, sold and advertised as a food product to aid laxation, but it has been introduced as a drug.

The conclusion I draw from all of this is that the system is somewhat confused. I fully understand the role of the FDA. I think they have the responsibility to decide what is safe for consumers in this country, a vital role, and I have respect for Dr. Kessler and the leadership that he is bringing to that institution.

In conclusion, one of the best long-term solutions to the spiraling health care costs in this country and around the world is disease prevention through improved diet and an exercise regimen. The food industry has enormous knowledge and the capability to develop the products, and we have enormous capability to educate consumers through truthful health messages in advertising. But if we don't get out of this regulatory paralysis, if things don't change, I believe the investment in research in improved foods is going to shut down in the food industry. The product development work will shut down too because there just isn't enough return on the investment.

This is a global issue because a lot of these products are developed in this country for expansion throughout the world. We draw this distinction between health claims and health messages. We can have debate about that, but I think the food industry in this country has been incredibly responsible. I think it will continue to be, and that it can manage this issue without much more in the way of additional regulations.

DISCUSSION PART VI:

MR. VAN ZWANENBERG: Some of these discussions have reminded me of a famous scientist who crossed an Idaho potato with a sponge. It tasted horrible, but it held a lot of gravy. The gravy people were very keen that we should sell it, as were the potato growers. But we weren't sure. It is difficult for a retailer to make up its mind, but we at least have a choice in our business, whether we sell a product or we don't sell a product.

The advice we take comes from many sources. Fortunately in the United Kingdom, we have a Food Advisory Committee, a committee of independent people, who are appointed by the government to advise ministers on food legislation. It is a non-statutory body. It consists of individuals who, it is hoped, have no axe to grind, and come from a wide range of activities, including science, consumer affairs, retailing, and production. They advise the government—very much in the way that the committee was suggesting here—on food legislation, and I think it is something that is worth considering for the United States. The United States is rather focused on the United States, and there is quite a lot to be gained by a greater dialogue between European food scientists and technologists and those in the United States.

There are a number of barriers to trade that are developing. Some of them have been already identified, but some of them are hidden. For example, the latest proposals in the United States for nutritional labeling are the direct opposite of the sort of nutritional labeling already agreed on in Europe. This is, in fact, a hidden barrier to potential trade between us.

In Europe, environment is a major public issue. It is going to affect, without doubt, trade. The proposed European directive on waste management is potentially going to put an onus on retailers and others to take back all the packaging that they sell. Already in Germany there is the Tofler law, coming into effect soon, whereby the industry in its totality has to manage the packaging that it sells so that it can, in fact, take it back.

All of these issues are going to affect trade. I think there should be more debate. There are potentially a lot of people who have interests. I don't think that one should ignore consumerism, and one has to listen to these debates. Whether they should take place in public or in private, I am not sure, but there should be more discussion in total. We talked about it before. The food chain is a total chain, and you can't ignore any part of it if we are going to be, as we hope to be, successful and trusted by our customers.

DR. MILLER: I believe that the public should be permitted to regulate or ban something because they just don't want it. In the United States, there is a procedure incorporated in the regulatory process to do so. If the public really doesn't want something, they have a forum through their elected representatives to see that it is banned.

Because of the nature of the U.S. Food and Drug law, FDA was compelled to begin a process to ban saccharin. The Congress, hearing from their constituents, said "no." And saccharin still remains on the market today. For those of you who think that is good because you may get more products on the market, that decision was fine. You have to understand, however, that the law has to be symmetrical. What is right for one side is also right for the other. The opportunity for the public to express their views must be available. I believe the system we now have in the U.S. works well. The elected representatives, not those that are self-selected representatives, are the ones who should make the determination that the decisions of the regulatory agencies are to be overturned.

Another point I want to make concerns health messages. The Food, Drug, and Cosmetic Act is carefully drawn to make a distinction between foods and medicines; a very strict, very careful distinction. We have now come into a world in which that distinction has rapidly blurred. I don't argue with the change. But the law has not kept up with it. Over the years, the FDA utilized the clear, legal distinction between foods and drugs as a tool for inexpensively dealing with food fraud. All the FDA had to do was to charge that a drug claim was made for this food. The courts would then declare that the validity of the claim was not germane since the issue was the process of drug approval. In most cases, the courts ruled that this was an unapproved new drug and demanded that the product be removed from the market at once.

In the years I was with the agency, we took many such actions, very inexpensively. The minute the concept of that distinction was challenged—first by Kellogg and then by other parts of the food industry, by the Congress, by the Administration—the FDA could no longer use this tool. The agency lost several cases trying to use the argument of drug claims as a tool to remove products from the market. The FDA was unable to prevail because the judge said, if you don't take action on those products making health claims, why are you taking action on the products in question? The agency now must demonstrate that a product is a fraud, which is a very expensive proposition.

Don't misunderstand me. I am not opposed to health claims. In fact, the original idea of health claims on food was made in testimony given by me and Mark Novich in 1980. The trouble was that we never were able to figure out how to accomplish it without producing the domino effect of taking an action that has implications much greater than anyone predicted. What I am really arguing is that if the capacity or need of the industry has exceeded the capacity of the law, it is time the law is changed. This takes place through a well-known process in the Congress. I would say to Representative Brown and his colleagues that complaining about how FDA administers the law doesn't do any good if FDA doesn't have the tools to enforce the law.

Internationally, there are two issues involved, and one involves the Congress again. I am not certain that the Congress, in considering changes in the food law, are taking seriously the question of harmonization with our other trading partners. Forgetting about the European Community, I am talking about our own discussions with the Canadians and the Mexicans for free trade. The National Labeling and Education Act, unfortunately, is causing concern in Canada, as it is in the European Community. Our colleagues in the Canadian government are certain that this is going to be a major sticking point in the negotiations. I think we have to consider these issues now. We never had to do so before.

From the point of view of technology, scientists in the U.K. scientific advisory committee for foods, scientists at MAFF, and scientists in virtually all of the regulatory bodies in the developed world, are in almost daily contact. By the time the issue arrives at the door of the EC Committee on Foods, they know all about it. They have all the data and have heard all the discussions. There is an international community of the regulatory scientists and they stay in touch with one another on a regular basis.

MR. HESS: It is my understanding that this Sunday "60 Minutes" will have a segment on the flavor enhancer MSG. To my knowledge, this is an approved safe food ingredient in most countries in the world, and I would be interested in what the opinion of this group is on the impact that will have, not only in the food industry in the U.S., but longer term on a global basis if we continue to have, for whatever reason, this type of trial by "60 Minutes."

MR. UTADA/TRANSLATOR: Our company manufactures and distributes food items and amino acids, including MSG and aspartame.

It is 80 years since we started to manufacture MSG. Various experts and scholars around the world have been spending many years conducting safety research, and the safety has been endorsed and recognized by various administrative organizations, such as the U.S. FDA, and United Nations organizations, such as WHO and FAO. We firmly believe that we can be proud of our activities for MSG, because MSG has been contributing to the dietary life and even dietary culture of the people around the world. They have used it not only in the food and processing industries alone, but also in private homes.

However, in any country there are a handful of people who tend to be opposed to this kind of thing, and once they raise the opposition, the mass media tend to amplify their voice. And consumer organizations sometimes are opposed to something, and that is, again, emphasized disproportionately by the press and the mass media. So, our company always has been fighting this tendency. The key point is how to get a good balance among technological progress, cultural progress, and the opinions raised by a handful of opponents to some new ideas and new technologies.

We must really give full trust to the power of science. However, the corporations should not be self-content, but rather actively work on consumers and the mass media to try to seek their understanding of our message. For this purpose, one-way publicity and advertising is not enough. We have to create opportunities to have effective dialogue with consumers and the mass media as well. The trust in the corporation's name is important. We have to create a corporate image that is reliable, that inspires confidence.

MR. WOODHOUSE: Obviously, this country doesn't know how to deal with the problem of certain individuals having allergies. An allergy, whether it be to sulfites or MSG, ought to be easily handled when it can be seen that there is only a minority of people around who are affected by it. Industry, by labeling, should allow the individual to make the choice whether he or she wants to use the product or not.

MR. BLOBAUM: There should be consideration of people who have problems with some of these products. And it is not that consumer organizations are opposing the products. They are just saying that there should be an opportunity for people to make the choice.

MR. HIRSCH: There are vast numbers of the American public who are in treatment with drugs that demand nutritional alteration, for example, monoaminioxidase inhibitors for the treatment of depression. With the growth of medical technology, we are going to find that more and more people will have very specific and different kinds of nutritional needs. Thus, there are individuals on such drugs who cannot eat cheese, chocolate, or other foods containing tyramine, without unfortunate adverse reactions.

This brings up the question of specific diets for people, and it relates to such recommendations as fiber intake. There may be some people who can profit by increased fiber intake and some who cannot. We are now beginning to have the techniques to find who can profit. There are better colorectal markers for malignancies. To the best of my knowledge, there have been no studies on humans of the effect of fiber intake under controlled clinical circumstances on essential aspects of colonic epithelial metabolism that bear on the proclivity to develop neoplasms. I would urge the food industries to be a part of this new clinical nutrition venture as you are a part of the bioengineering of food. The history of the food industry has in the past been a poor one in the support of fundamental clinical nutritional research.

DR. ARNTZEN: I agree very strongly that there is a real need for expanded, scientifically based clinical studies. Is the "60 Minutes" presentation on MSG going to have positive or negative effects, and is it going to promulgate any further public judgment of food safety? I would be interested in other people's perceptions of the depth of the public concern about food safety. In my own view, it is relatively limited to the few individuals who consistently refer to pesticide contamination as a starting point and then refer to all other things as further examples of major problems. I am sorry Ellen Haas left early, but I believe yesterday when she was sitting up at that table she mentioned pesticides 10 or 12 times. The references had no relevance to any of the other conversations that were going on, other than the fact that pesticide references are the mechanism that one can use to wake up the general public. Once you have talked about pesticides, then they will be conditioned to accept any other statements about dangers in our food supply.

I think as more things start coming out on "60 Minutes," *USA Today*, or other public communication channels, the public is going to be able to relate the topics to their own life. They are going to have known someone who has suffered from MSG difficulties and know that they didn't die, or they didn't get cancer from it, or they didn't suffer some other terrible problem. They are going to put it in the context of a risk and that it is an acceptable risk level. It is very positive to get that sort of thing out on "60 Minutes," because it is going to make people think that there are risks that they can tolerate and their neighbors are tolerating. They will begin to distinguish between dangers. The public would be served and the industry would be served if there were more debate on some things for which the public has experience.

DR. VAN DER HEIJDEN: It was about two years ago that the EC Scientific Committee on Foods was requested to present an opinion on MSG. We evaluated and discussed the extensive literature and we were completely unable to find any reliable study that demonstrated human effects. Maybe they are there, but I doubt it.

In this respect, I would make a strong plea for better studies. We have all heard those rumors from somebody. It would be very useful, just as in the pharmaceutical industry, to have a postmarketing surveillance for this type of material, some institutions where consumer complaints can go to, as has been done with the complaints about aspartame. The FDA did very good work to evaluate those data.

MR. BLOBAUM: If people get sick from MSG, they know it. This is true of someone who has intolerance to lactose or intolerance to wheat products or whatever. Labeling and disclosure are the issue, and it seems to me that it is up to the food industry to make it possible for those people to avoid those things that they know are going to make them ill.

PROFESSOR DUNLOP: This discussion reflected the changing nature of the food chain, its international elements in trade, its dynamic nature as a result of technological change and numerous changing fashions, and the question of whether the FDA of the past will be appropriate to the future. All of these questions illustrate a certain dynamism that underlies the questions of what ought to be our next steps.

PART VII

The Emerging Global Food System: Conclusions

Gerald E. Gaull and Ray A. Goldberg

The second Ceres Conference brought together leaders of the global food system and leaders of the public sectors affecting its operation in order to expand the discussion of food and nutrition policy begun two years earlier. Scientists, nutritionists, regulators, public policy makers, farm cooperative leaders, marketers, consumer groups, and academicians all were represented. They provided a global perspective of the revolutions in technology, information, resources, trade, economic development, and nutrition. The growing complexity and interrelationships of the world food system—including such phenomena as the advent of genetically engineered foods and the change in many countries from a centrally planned food system to one based upon market economics—have led to much debate concerning the use of appropriate technologies and the development of appropriate food safety and nutritional standards. The task for the future is managing this rapidly changing food system in such a manner that all players feel they have a voice in the provision of safe and nutritious food in sufficient quantity.

At this conference, discussion of management of the global food system fell into three major categories: (1) the effect of new technologies, including bioengineering, upon food production and distribution; (2) regulatory and legal issues, including trade and nontariff trade barriers; and (3) risk assessment and perception of risk by the consumer. What follows are inferences and conclusions from these proceedings.

FOOD PRODUCTION AND DISTRIBUTION

Information Technology

One central theme concerning production and distribution in the global food system provided possibility for consensus: The connections of the food chain are moving closer because of not only our increasing ability to custom-manufacture foods, but also the information technology that allows producers and distributors to determine more precisely consumers' wishes. Farmers are often asked to tailor their crop selection to the specific needs of producers and processors, who are pressed to change product composition or quality swiftly to meet the demands of a fickle marketplace. The computer age, as Thomas Urban suggested, "is driving information from the consumer back through the producer distribution system to the farm . . . we can almost identify certain farmers with certain shelf space."

Bioengineering

The techniques of genetic engineering abet our ability to custom-produce foods for a fragmented consumer market, and in turn have helped create the market itself. Many productivity gains in the recent past have come from classical genetics (cross-breeding, etc.) and traditional techniques will continue to contribute to the enhancement of food production. Engineered foods, however, will soon comprise a substantial portion of food products, and the new technology holds promise of improving yields and quality while decreasing development time. Currently field trials of over a dozen engineered crops are being conducted in many states, and there will be a burgeoning of crop testing in the next few years. Biotechnology may even offer possibilities for progress exceeding those of the Green Revolution.

There are several principal concerns about bioengineering that have been voiced by government and consumers. One of the most frequently mentioned is the question of the safety of releasing engineered foods into the environment. While scientists and industry must allow for the need to conduct careful environmental risk assessment, there are persuasive arguments that, far from being a menace to overrun the world, a bioengineered food would quietly die out if not carefully cultivated. Biotech products, rather than being able to "take over" in the environment, are almost always engineered for a special niche or characteristic, making them less able to survive in the evolutionary wilds.

A second issue is apprehension among farmers that bioengineering will create high-input-cost specialty crops and animals that only large-scale operations could economically produce. In this regard, the development of co-ops, such as the organization of thousands of dairy farmers into the Amul Dairy Co-op in India, has allowed small-scale farmers to share both the benefits and risks of investing in newly created crops and animal technologies. Additionally, biotechnology has the promise of reducing, not increasing, the cost of food production, as in the development of pest-resistant crops that will obviate the need for repeated spraying of pesticides.

A third concern involves the difficulty genetic engineering companies are having in getting actual products to market. Research, development, and testing of a product can take a decade or longer, a considerably long period in the volatile consumer—and investor—arenas. To facilitate long-term financing, biotech firms increasingly are working to expand formal arrangements with production and distribution companies, not only for research and development but for providing partial or whole foods downstream in the food supply.

Such a trend would fit with the overall pattern of increasing interdependence of all sectors of the food industry. A 1990 report of the U.S. Council on Competitiveness Biotech Working Group stated that government should "provide needed support in limited circumstances to activities that offer large potential benefits to the economy as a whole, but do not offer the prospect of adequate profit to any particular firm that might undertake it." Government, industry, and research companies must work together to ensure that bioengineered foods are nurtured through the lengthy development process, for bioengineering can enable a reliable and abundant food supply.

Distribution

Gerald Trant noted that "from 1966 to the present the world has been producing enough food to meet the daily caloric requirements of everybody. . . . This is a formidable record, brought about largely through yield increases, which have gone from 1.2 tons of grains equivalent per hectare in 1951 to about 2.3 per hectare presently"—on only 3% of the surface of the world. This achievement has been obtained through the use of chemical fertilizers, pesticides and herbicides, better crop management practices, growing farm mechanization, and development of high-yield wheat, rice, and maize. And yet, the number of chronically undernourished people in the world has risen from 450 million in the early 1970s to 550 million today.

Why are we unable to provide many of the world's people with adequate food, especially in light of the remarkable increases in productivity? The answer seems to lie less in food production than in our inability to distribute the food where it is needed at a price that is affordable, especially in less developed nations. There are two causes for this inefficiency in the food system, the first of which is geographic. Even with the growing internationalization of the food production system, much of the world's food supply is still produced in the wealthy, technology-wise areas of the world, instead of in the developing areas. Indeed, in Latin America food production has stagnated and in Africa there has been a decline. Second, the world distribution system has been stressed by the momentous changes in Russia and Eastern Europe, areas in which the movement of produce to market, never an efficient process, has in some cases almost stalled. The United States and Western Europe have pitched in, at least for the short term, by shipping large amounts of produce to these markets, a remarkable humanitarian gesture, but one which, unfortunately, may put a cap on what can be sent to Africa, Asia, and other areas of widespread starvation and malnutrition.

Another problem with distribution lies in national price supports, which tend to breed inefficiency in agricultural markets. Artificial price levels also hinder production of needed products in developing countries, which typically cannot offer such support to their farmers. The consensus at the Ceres Conference was that a freer market food system, without national supports that distort normal trade patterns, would in the long term do a better job of providing food where it is needed worldwide than does the current system. Robert Jackson suggested that "we need fundamentally to redesign agricultural support, shifting the focus from sustaining overall prices to sustaining selected producers and selected patterns of cultivation."

Rapid removal of price supports and other hindrances to an unfettered food system, however, would cause problems of severe unemployment and industry displacement. For example, in Poland, after years of over-regulation, centralized planning, and supports, there have been paradigm shifts toward barrier-free production and importing, with the result that inflation and budget deficits have been dramatically reduced. Yet, these moves have led to the underemployment or unemployment of thousands of farm and food production workers, and the Polish economy has nowhere to put them.

What is good for the world food system in the long run, then, is often disruptive to national systems in the short term, especially in Eastern Europe and other developing areas where up to 25% of the population is involved in small-scale farming. There is clearly a need for economic safety nets during the period of adjustment from national protectionism to international free trade, and the pace and extensiveness of such adjustments must take account of social and political realities in each country.

During the "shakedown" period, the developed countries should try to render substantial aid in the form of technical expertise, loans, and even direct food shipments to those countries whose food supplies and economies are most affected. Several measures to assist developing nations in creating free markets are currently being taken by the United States and other countries: an increase in the number of direct contracts and projects with multinational firms; a push for the reduction of Western barriers to agricultural imports from developing countries (it works both ways!), which is a central feature of the U.S. Trade Enhancement Initiative; provision of training and technical assistance; making available funds for assistance in privatization; recruitment of firms and academic institutions willing to send experts to developing countries; and encouraging the establishment of cooperatives, which allow small farmers to share the benefits and risks of reduced subsidies and modernization. Moreover, the larger international companies should consider providing advice and information to countries struggling to adapt to the new global food system. Such assistance might even be part of formal agreements between companies and governments, as part of the price of doing business in a country. These various measures ultimately will help the developing world move from being the object of humanitarian aid to being members of a more robust economic community, with meaningful and efficient markets.

REGULATORY AND LEGAL ISSUES

A pressing issue in management of the world food system is that different countries often have different approaches to protecting the safety and availability of their food supply. For example, there is much disagreement among policy makers concerning the definition of "safe" food. The United States and Europe tend to view safety regulation from different philosophical perspectives, and within both Europe and the United States there are numerous issues on which there is little consensus.

Safety Regulation in the United States

The United States food regulatory system is largely science-based: The goal is to evaluate the safety of a food or agricultural product, and if that product is safe for human consumption, to allow it on the marketplace. By extension, the final product, and not the process by which it is produced, is regulated—meaning a bioengineered food is to be treated like any other.

Regulatory procedure in the United States is largely transparent. Especially since the passage of the Freedom of Information Act, most government decisions concerning food and nutrition receive intense scrutiny by the media and consumer advocacy groups. In addition to this transparency, the original Administrative Procedures Act set up what is literally an adversarial process that lends itself to what can be a distressingly slow and acrimonious series of petitions, hearing, and lawsuits. The admirable intention was to allow interested parties to participate in an open regulatory system; unfortunately, the result too often has been to cripple that system.

Beyond the actual procedure of introducing regulations, the largest problem with American regulation of food safety lies in defining the term "safe." The question, and it is one on which American food regulations vacillate, is whether to employ a reasonable-risk standard, proclaiming a food or food additive safe when it is shown that there is only a very small risk of illness or death (e.g., 1 in a million), or, alternatively, to declare unsafe any food or additive that has been shown to increase the chance of illness or death, no matter how small that increase. The latter is called the zero-risk option.

The U.S. food regulatory system, divided among the Food and Drug Administration, the Department of Agriculture, and the Environmental Protection Agency, largely employs the reasonable-risk option, especially for approval of raw produce. However, in 1958 the U.S. Congress passed the Delaney Clause, which prevents the approval of any processed food containing a substance known to cause cancer in an animal, regardless of the level of risk of induction. This has led to what Roger Porter referred to as the "Delaney Paradox," in which raw foods that have been approved for consumption in the United States may become illegal as soon as they are processed.

Today this situation has been exacerbated by our vastly improved ability to detect minute quantities in food of ostensibly cancer-causing substances. The results are sometimes ludicrous, as in the recent past when the USDA had to ban the import of many European red wines because of the detection of insignificant levels of procymidone. An additional problem arises in the review of pesticides and herbicides that were approved for use before 1958. The USDA is being forced in some situations to ban the use of plant treatment substances that have been safely used for years, leading to the loss of these substances to agriculture and perhaps a resulting decline of productivity. There obviously can be no replacement with new substances of similar composition, as any new product would also be subject to the Delaney zero-risk limitation.

To deal with such problems, the 1989 President's Food Safety Plan made several recommendations: that the focus of food regulation should be on the product, not the production process; that regulatory burdens should be minimized while assuring public health and welfare; and that minimal reasonable risk, as opposed to zero risk, should be the safety standard. Additionally, the Office of Technology Assessment has proposed that there be two levels in the food approval process: a low-level review for new foods functionally identical or similar to existing foods and a higher review for truly original or unique foods.

Safety Regulation in Europe

The EC food regulatory system is divided among several bodies: the Commission, which initiates and administers legislation; the Parliament, which debates draft legislation, often voicing consumer concerns; and the Council of Ministers, which adopts legislation behind closed doors. The national food systems exhibit vast differences in the extent of food regulation, with the northern European countries generally placing more emphasis on food safety than those in the south. The EC has, however, been taking steps to harmonize regulations across the continent. After 1992 there is supposed to be a common food policy laying down some safety requirements and facilitating removal of internal barriers to trade, though food composition laws have been put on hold. Further, the role of the EC Scientific Committee for Food is being expanded: Among other functions, the Committee is supposed to allocate different foods and substances to different countries for testing, in order to avoid duplication of efforts.

Even as common legislation is passed, responsibility for enforcement remains with member nations, a cause for concern among those countries with greater regulatory emphasis (or at least larger regulatory budgets). This has led some to call for greater enforcement power for EC agencies, or even the creation of a European FDA.

Green parties are well established in the European legislative process, so that consumer and environmental concerns are often a major force in the passage of laws. One resulting difference between the American and European food regulation systems is that in Europe the process by which a food or substance is produced is seen as falling under the scope of regulation. Thus, a product is much more likely to be rejected for use because of being bioengineered than would be the case in the United States.

A second difference is that in many of the EC countries, a "need" assessment is built into the formal approval process. That is, in addition to a finding that a substance or food is safe for human consumption (the only initial requirement for American approval), there must be a determination that the introduction of a new crop or food would fit with the overall economic and cultural patterns of living. A food might be rejected if, for example, local farmers would be put out of work, or if a native product would lose shelf space. This process also occurs in the United States, but through other means, principally consumer and producer advocacy and media pressure; the point is that need or social valence of the product is not a built-in component of American food regulation.

An interesting and somewhat discouraging commonality between the United States and Europe is that neither of the systems, as complicated and exhaustive as they are, cover foods intended for export from their jurisdictions. Eliminating this seeming disparity in concern for other populations will be a desirable part of the future global food system.

International Food Trade and Regulation

The Uruguay round of GATT talks focused in large part upon the problem of bringing national food trade policies and safety regulations into closer agreement. There were several key points of discussion: strengthening of rules on exports, support policies, and import restrictions; development of rules to protect international property rights; basing food regulations on scientific evidence of safety (as opposed to social need); and an allowance for countries to set standards higher than there called for by international regulations.

The proposed mechanism of regulatory harmonization would be adopting and strengthening the language of the Codex Alimentarius, a United Nations agreement to which 130 countries have been signatories. Codex is perhaps an admirable first step toward harmonization of world food standards, but it is a first step only, for it would seem that the goals of basing regulations on scientific evidence and yet allowing for high national safety standards might be subverted by parochial interests. The concern here is that a national food safety regulation more stringent than the level judged appropriate by international scientists and regulators might represent back-door implementation of a nontariff trade barrier. Without denying the legitimacy of cultural and national preferences in lifestyle and dietary habit, there must be a balance between such concerns and the need for an efficient global food system.

A second fundamental problem is that scientific expertise and resources vary greatly among Codex nations. In effect, countries with a meager scientific infrastructure are forced to depend upon the decisions of those few countries that have the expertise and can afford extensive testing and approval programs. Because of this discrepancy and the allowance of high national standards, Codex would seem to allow the developed countries to do as they will, while developing countries in many cases would endure hardship were they required to meet higher standards. First, such countries would be forced to allocate scarce funds to increased food testing programs. Second, food that is produced or processed to such standards would be more expensive than that currently imported to or produced in many developing areas. While food safety is certainly a legitimate goal everywhere, simply providing enough food for the populace is often the overriding concern for developing nations.

There may never be a completely free market with full agreement among nations concerning regulation of food. There are cultural differences, disagreements concerning the needed level of safety, and considerations of who might suffer unduly if economic safety nets were to be eliminated. At the same time, the global food system needs rules concerning food safety, product standards, nutritional requirements, and trading procedures, and all nations should continue to work toward agreement on these matters.

FOOD SAFETY ASSESSMENT AND CONSUMER PERCEPTION

The nature of science is such that there is inevitably some degree of uncertainty in research data, and thus corresponding uncertainty in their interpretation. Thus, debate among scientists concerning research is to be expected and tends to be iterative, leading to further hypotheses and testing. The public, however, does not understand that no responsible scientist ever insists that he has the "final answer" to a research question. While it might be true that scientific issues in nutrition and food safety often can be resolved when experts meet, were the same panel of experts to be reconvened in several years, it is entirely possible (and is often the case) that the results would be different. Further, as Thomas Kuhn, the philosopher of science, pointed out several decades ago, scientists and the supposedly neutral process of "pure" science are far from immune to cultural conventions and expectations.

Even when scientists can predict with precision what will obtain over the near term in their laboratories, when it comes to seeing into the future they are often as fallible as nonscientists. Two decades ago molecular biologists were candid in informing the public of what they thought were the possible hazards of the new science of genetic engineering, and as a group they worked with the National Institutes of Health to create very restrictive guidelines. There is now consensus among scientists that most of the early fears were unfounded, and the guidelines have been relaxed.

Such, too, has been the history of the effort to control food additives. Following World War II there came to be general scientific agreement that many foods and food additives contained suspected carcinogenic agents. The presence of at least some of the agents could be detected, and regulatory effort in the United States came (and continues) to be directed to controlling unnecessary exposure to these substances. In the past decade, however, the importance of direct carcinogens in food has come to be questioned as the number of alleged agents has grown by scores, and it is realized that we are unavoidably exposed every day to many of them. If the hazards are so great, why do we not all have cancer?

As a scientific discipline risk assessment is yet nascent, having shown signs of maturing only recently. This, in combination with the inherent ambiguity of science, should lead us to view with caution any claims that a risk, particularly long-term, associated with a food or food additive, can be determined with certainty. Unfortunately, this fact fuels the argument for both sides of the food safety issue. Those who would minimize government's intervention in approving new foods can point to numerous cases in which dire predictions of human or environmental harm have been wrong. Others believe that our inability to pinpoint the nature and extent of risks from food is the very reason that we should choose to err on the side of caution, with extensive regulation of products until time has shown that they are indeed safe.

One problem in both the development of risk assessment and in the perception by many of its usefulness is that it is often used to try to validate a negative proposition. No matter how long we test a food or how diligently we extrapolate into the future, there simply cannot be a guarantee that no harm will ever come to anyone. The search for such elusive and chimeric assurance is, unfortunately, in danger of becoming a hallmark of US food safety efforts. Such a quest can be undertaken only at great economic expenditure and with minimal potential return.

Risk Communication and Public Perception

People around the world (though perhaps mostly in the developed countries) are becoming increasingly concerned about what they eat. As Alan Bromley pointed out:

"America has the safest, best, and most abundant food supply in the world. Yet public concern about the quality of that food supply is extremely high—and seems to be going up as our food improves!". Consumers demand that they be provided with "scientific evidence" (albeit in sound-bite doses) about the benefits and risks of the thousands of foods and food products found in the supermarket. Such skittishness is perhaps only to be expected, given the simultaneous trends toward educating consumers to adopt healthful diets and toward providing public warnings whenever a suspected or alleged food health risk is identified.

The public as a whole is not scientifically literate, and it is aware of that—which, paradoxically, is precisely why it demands that science tell it what to do. In these times of bioengineered foods, food additives, pesticides, and herbicides (known generically to the public as "chemicals"), and a seemingly never-ending number of substances "found to have a link to cancer," terms such as "Frankenfoods" capture the public imagination and fuel fears of unknown science. Consumers do deserve to have their fears addressed in the media by responsible scientists. Unfortunately, in some cases it is the very fact of scientists opening up the research and risk assessment process that makes the public so nervous. What, then, should be the extent and role of risk communication? Dorothy Nelkin summarized the problem well: "How can messages be designed that do not oversimplify the scientific uncertainties, yet are still appropriate for publication?".

A related question involves the responsibility of corporate use of risk assessment data. There have been recent cases in which many scientists felt benefits were being claimed that were unproven or exaggerated, as, for example, in the use of epidemiological data linking fiber intake with a reduced rate of cancer in marketing breakfast cereal.

While there is much room for debate concerning the particular criteria for and appropriate timing of scientific disclosure, academic and industry scientists should (and do) strive to present responsible assessments of food health benefits and risks to the public, so that consumers can make informed choices about what they eat. There is, however, an important caveat. In an open society, it is imperative that citizens be educated about the scientific process itself and in particular the uncertain nature of research, including health risk assessment. As Ellen Haas noted, "communicating risk is itself a risky business," made all the riskier when the recipients of the communication are unprepared to interpret it.

Education of consumers should also include information about the expense to themselves of providing zero-risk food. Science and technology can do much to improve the safety and quality of the foods we eat, and it can warn us of the presence of any number of possible health risks—and government can certainly move to restrict unsafe foods from the marketplace. Such security, however, comes at a cost: the cost added to production in order to obtain zero risk and the lost opportunity costs incurred by diverting resources. With so much public concern about bioengineered foods and the presence of carcinogens and other hazardous agents in food, it is possible and perhaps even probable that legislators in the Uned States and other developed countries will feel compelled to spend large amounts of public funds in an effort to create an ostensibly zero-risk environment—money that could perhaps be better spent to work for higher agricultural output of foods falling within reasonable risk guidelines. If the public truly wants zero risk, then perhaps they should get it—but they also should be made aware that it does not come free.

IMPLICATIONS FOR THE FUTURE: COMING TO COMITY

Throughout this second Ceres Conference a number of specific suggestions were set forth concerning the roles of the various participants in the emerging global food system (see Tables 1-4).

The principle task, however, for public officials, food producers and manufacturers, scientists, and public advocacy groups is to find ways to work together in better serving their roles. Only by focusing on this task can we fulfill the goal of creating an efficient yet safe global food system.

The major challenge in achieving this goal is to alleviate the misunderstandings and stridency of debate that often characterize current discussions of food and nutrition policy and research—and to strive for comity. The various sectors of the food system frequently find themselves in bitter adversarial relationships, when at most the data allow for vigorous interpretive disagreement. Further, adversarial attitudes occur not only among groups but among individuals within a group: Researchers debate the relative health risk of foods and food additives; regulators argue about whether to use zero-risk or reasonable-risk criteria; manufacturers look for a free market, but are angered when ill consequences (at least in the short term) are seen for their particular industry; consumer advocates differ on whether protection of the public should emphasize strict government control or consumer education.

The difficulty in coming to agreement on such issues, however, goes beyond differing opinions on scientific evidence or regulatory priorities. Increasingly, the public seems to be unsure if its elected and appointed officials really are working for the good of the people. Nor are others free from suspicion. Corporations are often seen as working for profit—without regard to environmental or public health risks. Consumer advocate groups are suspected of placing more importance on grantsmanship and their own survival than on reasonably addressing issues. Scientists are accused of conducting dangerous research with little thought about the consequences, and increasing collaboration with industry leads to charges of scientists having abandoned their disinterestedness.

This public distrust of government and large institutions is not unique to the food industry, nor is it of recent origin. Particularly in the United States, such wariness is long-standing. When, however, the matter at hand is as basic and important as what one eats and its relation to one's health, concerns about the trustworthiness of establishment institutions are intensified.

This lack of confidence in institutions, moreover, is not characteristic just of the public, but is sometimes shared by those within the food system: Corporations tend to view government regulation as stifling innovation. Policy makers wonder whether scientists and industry leaders provide them with accurate information about new foods and technologies. Scientists feel that they get blamed for not being able to foresee all the consequences, good and bad, of their research. And all of this mistrust is grist for the mill of the media, who are under their own pressures of time, resources, professional ambition, and, yes, sale of their product—and whose opinions (as is the case with all others) may be shaded by ideological bias.

It is imperative that we find a way around group animosities and inevitable scientific uncertainties to do a better job of managing the global food system. The primary value of the Ceres Conferences lies in bringing together in a single room leaders of the various sectors of the system—scientists, producers, distributors, legislators, regulators, consumers, and advocacy groups—to sit down and discuss these issues. These leaders agreed that no single traditional discipline by itself can provide adequate solutions for the many problems we face. New models and cross-disciplinary approaches to food production, distribution, and regulation must be developed.

Finally, throughout this Conference there was a recurring theme: The present immediacy of communication concerning food benefits, risks, and regulations work to the detriment of the common good because of the confusion engendered by multiple and opposing messages. This will continue unless neutral organizations, blending expertise from multiple fields of inquiry, can be developed. The public and their translators, the media, then could turn to such groups for rational, reliable, and clear information or guidance concerning the safety of the food they eat and how it will affect their nutritional well-being and concerning the economic development of their global and national food systems.

Table 1 Role of Governments in the Emerging Global Food System

Science/Technology

- Encourage biotechnological innovation.

Production/Distribution

- Provide market orientation to centrally planned economies.
- Provide safety nets for transition to market economies.
- Reduce export subsidies and domestic price supports, and provide market access.
- Assist the private sector in assessing and meeting global training needs.

Regulation/Social Concerns

- Develop international patent and intellectual property legislation.
- Develop international health-related standards based on scientific evidence.
- Strengthen Codex Alimentarius or develop an acceptable alternative with responsibility to enforce international safety standards.
- Design national food safety legislation and a national framework for making assessments of acceptable risk.
- Strengthen existing health/regulatory agencies with additional qualified personnel.
- Abet development of new private-public forums to discuss initiatives for cooperating on trade, food, and technology policies and procedures.
- Develop forums to discuss the social and economic consequences of adopting new technology.
- Involve consumer groups in the development of policies for new technologies.

Table 2 Role of the Private Sector in the Emerging Global Food System

Science/Technology

- Use scientific advances in biotechnology to abet classical genetics in the improvement of foods.
- Develop continuing relationships with scientific, government, consumer, and producer groups to provide the public with a better understanding of biotechnology and its importance to society's goal of an efficient and environmentally sound food system.
- Coordinate the application of technological possibilities with consumer needs.

Production/Distribution

- Assess and assist in meeting global training needs.
- Build industrial global infrastructures that can successfully merge old and new technologies.
- Assist government in providing market orientation to centrally planned economies.

Regulation/Social Concerns

- Assist government in developing new private-public forums to discuss initiatives for cooperating on trade, food, and technology policies and procedures.

Table 3 Role of Scientists and Academic Researchers in the Emerging Global Food System

Science/Technology

- Continue basic research in molecular biology to provide the scientific underpinning for future academic, government, and private sector decisions concerning biotechnology.
- Provide bridges to hospitals, farms, and corporations as well as research laboratories to provide pilot-testing of new drugs, plants, animals, processes, genes, and antibodies.
- Form joint ventures with governments to study rain forests and other lands as sources for unique medicines and other flora-based substances.

Regulation/Social Concerns

- Work with government, industry, institutes, and consumer groups to develop priorities in meeting the global challenges of the environment, education, health, and economic development.
- Assist government in developing health-related standards based on scientific evidence.
- Assist government in designing food safety legislation and a national framework for making assessments of acceptable risk.
- Provide greater assistance to governments, the private sector, and public advocacy groups in explaining new technologies to the lay public in language that is readily understood without sacrificing accuracy of content.

Table 4 Role of Public Advocacy Groups in the Emerging Global Food System

Regulation/Social Concerns

- Provide the public with balanced information from governmental, scientific, and industrial sources concerning food and nutrition research and policy, translated into language that can be readily understood without sacrificing accuracy of content.
- Provide a voice for public interests and concerns about food and nutrition, both in informal forums with government, science, and industry leaders and in the formal regulatory process.
- Assist government in developing forums to discuss the social and economic consequences of adopting new technology.
- Explain to their constituents the likely costs as well as the proposed benefits of new policies and regulations.
- Join in forums and research with participants in the global food system from both the private and public sectors.

Biographies of Contributors

CAROL C. ADELMAN, D.P.H., Assistant Administrator for Europe, U.S. Agency for International Development. Previously Dr. Adelman was Assistant Administrator for Europe and the Near East of the U.S. Agency for International Development (AID). Now as head of the Bureau for Europe, Dr. Adelman manages U.S. foreign aid programs in Central and Eastern Europe. From 1970 to 1981 Dr. Adelman was a career Foreign Service officer with AID, developing projects in the Middle East after working in Africa on agricultural, health, and nutrition programs.

CHARLES J. ARNTZEN, Ph.D., Dean, College of Agriculture and Life Sciences, Deputy Chancellor for the Agricultural Program, Texas A&M University. Prior to joining Texas A&M, Dr. Arntzen held faculty positions at the University of Illinois and Michigan State University. He has served on the scientific advisory board for Advanced Genetic Sciences and Plant Genetics System in Brussels. He currently serves on the Board of Directors of DeKalb Genetics, Inc. and AgriStar, Inc. where he also is the Chief Scientific Advisor, on the Board of Governors of the University of Chicago for the Argonne National Laboratory, and on the Scientific Board of the Hong Kong Institute of Biotechnology. As a result of his pioneering research in photosynthesis and plant molecular biology, Dr. Arntzen was elected to the National Academy of Sciences in 1983.

ROGER BLOBAUM, Director, Americans for Safe Food, Center for Science in the Public Interest. Mr. Blobaum's professional background includes experience as an agriculture staff member in both houses of Congress, as director of public relations for two national farm organizations, and as president of a consulting firm specializing in designing and managing alternative energy and agriculture projects. He has served on the U.S. Department of Agriculture's Cost of Production Advisory Committee, the Department of Energy's Food Industry Advisory Committee, the Solar Energy Research Institute's National Advisory Committee, and on the Congressional Office of Technology Assessment Panels.

D. ALLAN BROMLEY, Ph.D., Assistant to the President for Science and Technology. Dr. Bromley is on leave from his former position as Henry Ford II Professor of Physics at Yale University, where he was founder and Director of the A.W. Wright Nuclear Structure Laboratory. Prior to his present appointment, Dr. Bromley served as a member of the White House Science Council during the Reagan administration and as a member of the National Science Board from 1988-89. He has received numerous honors and awards, including the National Medal of Science. As president of the American Association for the Advancement of Science and of the International Union of Pure and Applied Physics, he has been one of the leading spokesmen for U.S. science and for international scientific cooperation.

GEORGE E. BROWN, JR., Member of Congress. Mr. Brown has represented the Riverside-San Bernardino-Ontario area of California since 1972. He is a senior member of the Agriculture Committee, chairman of the Subcommittee on Department Operations, Research and Foreign Agriculture, and chairman of the House Science, Space and Technology Committee. Mr. Brown's primary legislative interest for many years has been science and technology policy.

JOHN J. COHRSSEN, J.D., Associate Director of the President's Council on Competitiveness, Office of the Vice President. Previously, Mr. Cohrssen served as Senior Advisor to the Chairman of the Council on Environmental Quality, Executive Office of the President; and as regulatory counsel to the Office of Science and Technology Policy, Executive Office of the President. He also served as legal counsel to the Domestic Policy Working Group on Biotechnology and the Biotechnology Science Coordinating Committee, and was a member of the Vice President's Council on Competitiveness Working Group on Biotechnology.

BARBARA J. CULLITON, Deputy Editor, *Nature*, Visiting Professor, Johns Hopkins University. Ms. Culliton has held previous editorial or correspondent positions at *Science*, *Medical World News*, and *Science News*. Ms. Culliton has served as president of both the National Association of Science Writers and the Council for the Advancement of Science Writing. She is an elected member of the Institute of Medicine, National Academy of Sciences and was awarded the George Polk Journalism Award.

BERNARD D. DAVIS, Ph.D., Adele Lehman Professor of Bacterial Physiology, Emeritus, Harvard Medical School. Professor Davis joined the faculty of Harvard Medical School in 1957, was named Adele Lehman Professor in 1962, and in 1968 became director of the Bacterial Physiology Unit. During his tenure at Harvard he also was a visiting Investigator at the Weizmann Institute, Rehovoth, Israel; a Fellow at the Center for Advanced Study of Behavioral Science; a Visiting Professor at Tel-Aviv University Medical School and at the National Taiwan University Medical School; and a Fogarty Scholar at the National Institutes of Health. In 1989 Professor Davis received the Waksman Medal of the National Academy of Science and the Hoechst-Roussel Award of the American Society of Microbiology. He recently edited *The Genetic Revolution: Scientific Prospects and Public Perceptions.*

JOHN T. DUNLOP, Ph.D., LaMont University Professor, Emeritus, Harvard University. Dr. Dunlop previously was Professor of Economics and Dean of the Faculty of Arts and Sciences at Harvard University. He has served every President since Roosevelt, including appointment as Secretary of Labor under Ford. His honors include the Murray, Meany, Green Award of the AFL-CIO, and nomination to the National Housing Hall of Fame. He is a member of the American Academy of Arts and Sciences, the American Philosophical Society, and the National Academy of Arbitrators.

GERALD E. GAULL, M.D., Director, Center for Food & Nutrition Policy and of the Ceres Forum, Georgetown University. Dr. Gaull was formerly Vice President for Nutritional Science, The NutraSweet Company. He was previously on the faculties of Harvard University, Columbia University, and Mt. Sinai School of Medicine. His academic career spans 25 years in research and teaching with emphasis on nutrition. Dr. Gaull's honors include the Borden Award in nutrition and the Gold Medal of St. Ambrosiano of the City of Milan.

RAY A. GOLDBERG, M.B.A., Ph.D., Moffett Professor of Agriculture and Business, Harvard University and head of the Agribusiness Program at the Harvard Graduate School of Business Administration. He serves as a director, trustee, or advisor to many business, governmental and nonprofit organizations such as Pioneer Hi-Bred International, Archer-Daniels-Midlands Co., John Hancock Agricultural Committee, Rabobank, Transgenic Sciences Inc., Eco-Science Inc., and Caribbean Basin Initiative. He is the founding president of the International Agribusiness Management Association and is chairman of the board and treasurer of the Agribusiness Institute of Cambridge. He also was a member of the Presidential Mission to Poland in December 1989.

ELLEN HAAS, Founding Director, Public Voice for Food and Health Policy, a national nonprofit research, education, and advocacy organization dedicated to making food safety, better health, and improved nutrition national priorities. Ms. Haas was elected to five terms as President of the Consumer Federation of America, the nation's largest consumer organization. She also served for seven years as Director of the Consumer Division of the Community Nutrition Institute, a nonprofit group dealing with hunger and food policy issues. She currently serves on the Board of Trustees for the Food and Drug Law Institute and as Vice President of the Consumer Federation of America.

JOHN HIGGINSON, M.D., Clinical Professor of Community Medicine, Georgetown University Medical Center. Dr. Higginson served as the founding Director of the International Agency for Research on Cancer (IACR) of the World Health Organization in Lyon, where he established its scientific program and developed its global orientation. Prior to joining the IACR, Dr. Higginson conducted research in Africa on the role of environment in chronic disease and was American Cancer Society Professor of Geographic Pathology at the University of Kansas. He is a Fellow of the Royal College of Physicians (London).

CAROLINE JACKSON, Ph.D., Member of the European Parliament. Dr. Jackson is the Conservative Party spokesperson in the European Parliament on consumer protection and environmental affairs. Prior to this, she was a member of the U.K. National Committee for European Environment Year, a member of the National Consumer Council, Chairman of the British section of the Foreign Affairs Commission of the European Union for Women, and head of the London office of the European Democratic Group (Conservative in the European Parliament).

ROBERT JACKSON, M.P. and Parliamentary Under-Secretary of State, Department of Employment. Mr. Jackson has responsibility for Training Strategy and Programmes, Education Programmes, the Employment Service, and Equal Opportunities. His previous appointment was as Parliamentary Under-Secretary of State at the Department of Education and Science. In 1979 Mr. Jackson was elected Member of the

European Parliament for Upper Thames, and in 1980 he served as Special Advisor to the Governor of Southern Rhodesia (Lord Soames).

ARTHUR KORNBERG, M.D., Professor Emeritus (active), Department of Biochemistry, Stanford University. Dr. Kornberg served as Professor of Biochemistry since 1959 and was Chairman of the department from 1959 to 1969. He previously served as Professor and Head of the Department of Microbiology at Washington University. He received the Nobel Prize in 1959 for his research on the enzymatic synthesis of nucleic acids.

ARNOLD G. LANGBO, Chairman and Chief Executive Officer, Kellogg Company. Since joining Kellogg in 1956, Mr. Langbo has held several positions, including President and Chief Operating Officer of Kellogg Company; President and Chief Executive Officer of Kellogg Salada Canada Ltd. Inc.; President, U.S. Food Products Division of Kellogg Company; Chairman and Chief Executive Officer, Mrs. Smith's Frozen Foods Co.; and President, Kellogg International.

SANFORD A. MILLER, Ph.D., Dean, Graduate School of Biomedical Sciences and Professor, Departments of Biochemistry and Medicine at the University of Texas Health Science Center at San Antonio. He previously served as Director of the FDA's Center for Food Safety and Applied Nutrition and as Professor of Nutritional Biochemistry at the Massachusetts Institute of Technology. Dr. Miller's awards include the Award of Merit of the Food and Drug Administration and the Conrad E. Elvejhem Award for Public Service of the American Institute of Nutrition.

DOROTHY NELKIN, University Professor, Professor of Sociology and Affiliated Professor, School of Law, New York University. Professor Nelkin's research focuses on controversial areas of science, technology, and medicine as a means to understand their social and political implications and the relationship of science to the public. She has served on the Board of Directors of the American Association for the Advancement of Science and on the National Academy of Sciences Committee on National Strategy for AIDS. She is on the board of Medicine in the Public Interest, and the Council for the Advancement of Science Writing. She has been a Guggenheim Fellow, a Visiting Scholar at the Russell Sage Foundation, and the Clare Boothe Luce Visiting Professor at NYU.

CHARLES R. NESSON, J.D., Professor of Law, Harvard Law School. Since joining the Harvard Law faculty in 1966, Mr. Nesson has served as Associate Dean, as Organizer and President of the Lawyers' Military Defense Committee, and as Director, Harvard Evidence Film Project. He has extensive experience in the media, including serving as moderator for two PBS series, "The Constitution: That Delicate Balance," and "Ethics in America," as well as for the CBS program "Eye on the Media: Media and Business." He also has been an advocate on the WGBH program "The Advocates," narrated the film "Three Appeals," and has been the commentator for the documentary "The Shooting of Big Man."

ANDRZEJ OLECHOWSKI, Ph.D., State Secretary, Ministry of Foreign Economic Cooperation, Poland. Prior to assuming this position, Dr. Olechowski was Deputy Governor, National Bank of Poland; Director of the Ministry of Foreign Economic Cooperation;

Director, World Bank Cooperation Bureau, National Bank of Poland; Advisor to the President, National Bank of Poland; Economist, International Economic Research Division, Development Research Department, The World Bank; Economics Affairs Officer, United Nations Conference on Trade and Development (UNCTAD); Head of the Department of Analysis and Projections, Foreign Trade Research Institute; and Associate Economic Affairs Officer, UNCTAD, Multilateral Trade Negotiations Project.

ROGER B. PORTER, Ph.D., Assistant to the President for Economic and Domestic Policy. Prior to this, Dr. Porter was IBM Professor of Government and Business at Harvard University and Faculty Chairman of the Program for Senior Managers in Government. During the Ford administration he was Special Assistant to the President and Executive Secretary of the President's Economic Policy Board. During the Reagan administration he served as Deputy Assistant to the President and Director of the White House Office of Policy Development. Dr. Porter was awarded a Rhodes Scholarship and was selected as one of the Ten Outstanding Young Men in America in 1981 by the United States Jaycees.

JEREMY RIFKIN, President, Foundation on Economic Trends, President, Greenhouse Crisis Foundation. Mr. Rifkin has been actively involved in environmental issues and science and technology policy for nearly two decades. He has written 11 books that concern a wide range of interdisciplinary topics from politics to philosophy and science. His commentary has been featured on numerous television programs and in the popular press. Mr. Rifkin also has testified before numerous congressional committees.

WILLIAM RUDER, Founding Partner, Ruder & Finn, Inc., Chairman of the Board, William Ruder Incorporated. Mr. Ruder has served as Assistant Secretary of Commerce under President Kennedy. Together with his partner, David Finn, Mr. Ruder is credited with developing many new techniques and concepts in public relations and publicity. He also initiated the Department of Commerce Business Ethics Advisory Council, a group of outstanding businessmen, philosophers, and academicians joined together to encourage the highest standards of ethical performance in business.

ROGER SALQUIST, Chairman and Chief Executive Officer, Calgene, Inc., a leading international agribusiness biotechnology company. Prior to this, Mr. Salquist served for six years as Chief Financial Officer of Zoecon Corporation, the original developer of biorational pesticides, and for two years he was President of the Turf and Garden Division of Occidental Petroleum. His outside activities include serving as Chairman of the California Industrial Biotechnology Association, and he is currently Chairman of the Agriculture Section of the International Biotechnology Association. He is a director of both Collagen Corporation and Celtrix Laboratories.

JOZEF S. SCHELL, Ph.D., Director, Department of Genetic Principles of Plant Breeding, Max Planck Institute. Prior to joining the Institute in 1978, he was a professor at Rijksuniversiteit Gent (Belgium) and at Free University Brussels. He also is an honorary professor at the University of Cologne. Dr. Schell serves on the editorial boards of numerous journals and is on the Board of Governors of The Weizmann Institute of Science (Israel) and The Hebrew University of Jerusalem. He is a member of the Deutsche Akademie der Naturforscher Leopoldina, the National Academy of Sciences (USA), and the Academia Europaea, London as well as many others. Dr. Schell's numerous awards include the "Mendel Medaille" of the Deutsche Akademie der Naturforscher Leopoldina, the IBM Europe Science and Technology Prize, the Wolf Prize in Agriculture, and the Australia Prize for Agriculture and Environment.

JONATHAN F. TAYLOR, Chief Executive Booker plc. Mr. Taylor has been involved in all aspects of Booker's international agribusiness operations, becoming chairman of its agriculture division in 1976. He was responsible for the development of Booker's interests in fish farming, farm and forest management, plant breeding, and major agricultural projects in developing countries. In 1980 Mr. Taylor became President of Ibec Inc., a corporation jointly owned by Booker and the Rockefeller family. Under Mr. Taylor's leadership Booker has grown to be the U.K.'s largest wholesaler to independent food retailers, the leading food service distributor to caterers, a major international force in fish and poultry genetics, as well as operating a number of food processing businesses.

HARRY E. TEASLEY, JR., President and Chief Executive Officer, Coca-Cola Nestle' Refreshments Company. Mr. Teasley has been with the Coca-Cola Company since 1961. Prior to assuming his current position, he was President and Chief Executive Officer of Coca-Cola Foods; Managing Director of the English Bottling Operations of the Coca-Cola Company in London; and President and Chief Executive Officer of the Wine Spectrum. He was made Vice President of Coca-Cola USA in 1973 and Vice President of The Coca-Cola Company in 1975. In 1991 Mr. Teasley was awarded the Rene Dubos Environmental Award.

GERALD I. TRANT, Ph.D., Executive Director (Assistant Secretary-General) United Nations World Food Council. Prior to assuming this position, Dr. Trant was Senior Assistant Deputy Minister of Agriculture Canada. He also served as Chairman of the Agricultural Stabilization Board and of the Agricultural Products Board, and as a Director of the Farm Credit Corporation. Dr. Trant was a member of the Canadian negotiating team during the Tokyo Round of Multilateral Trade Negotiations, and he was a member of the Technical Consultative Group on International Agricultural Research. He also served as Assistant Deputy Minister for Economics of Agriculture Canada.

RYUICHIRO TSUGAWA, Ph.D., Director of Pharmaceuticals Development, Board Member Ajinomoto Company, Inc. Dr. Tsugawa joined the Ajinomoto Company in 1957 and held several positions, including Manager of the Microbiological Chemistry Department and Deputy General Manager of the Central Research Laboratories. His current research involves development of biologically active substances, such as interleukin. Dr. Tsugawa is Chairman of both the U.S.-Japan Bilateral Biotechnology Forum and the international exchange committee of the Japan Bioindustry Association. He is also an expert member of the Science Council, Japan.

EARL UBELL, Health and Science Editor, WCBS-TV News. Prior to assuming his current position in 1966, Mr. Ubell was Senior Producer, special broadcasts, and Director, TV news for NBC News. He also was a syndicated columnist with the *New*

York Herald Tribune. Mr. Ubell has received numerous awards for journalistic achievement, and he has served as President of the Council for the Advancement of Science Writing, of the National Association of Science Writers, and of the Gamma chapter of Phi Beta Kappa.

THOMAS N. URBAN, Chairman and President, Pioneer Hi-Bred International, Inc. Mr. Urban joined Pioneer in 1960 and held several positions, including President and Chief Executive Officer, before assuming his current position in 1984. He also was Mayor of Des Moines from 1968 to 1971. He is on the Board of Directors of several organizations, including Pioneer Hi-Bred International, Bankers Trust Company, Equitable of Iowa, The Weitz Corporation, Health Policy Corporation of Iowa, and Sigma-Aldrich.

KEES A. VAN DER HEIJDEN, Ph.D., Assistant Director, WHO European Centre for Environment and Health. From 1968 to 1977 Dr. van der Heijden was employed by Duphar Chemical-Pharmaceutical Industries and by the Dutch Organization for Applied Research. In 1977 he joined the National Institute of Public Health and Environmental Protection in the Netherlands as a senior toxicologist. In subsequent years he became head of the Laboratory for Carcinogenicity and Mutagenicity, the Toxicology Advisory Centre and the Laboratory of Toxicology. In 1987 he was appointed Director of Toxicology, responsible for scientific research and advice to both national and international institutions. He currently is chairman of the EC Scientific Committee for Food.

HERMANN VON BERTRAB, Ph.D., Director, Office of the Free Trade Agreement Negotiations, Embassy of Mexico, Washington, D.C. Prior to his current position, Dr. von Bertrab was President of IMERVAL (Instituto del Mercado de Valores), a brokerage house in Mexico City. He also served as Senior Vice President of Operadora de Bolsa, an investment house in Mexico; coordinator of economic and social planning in the Secretariat of Planning and Budget; President of Grupo BECCSS; and Senior Vice President of Bancomer.

JOHN F. WOODHOUSE, Chairman and Chief Executive Officer, Sysco Corporation. Prior to joining Sysco as Chief Financial Officer in 1969, he served as Treasurer, Cooper Industries, Inc.; held various financial supervisory positions with Ford Motor Company; and was an investment analyst with Canadian Imperial Bank of Commerce. Mr. Woodhouse holds numerous director and trustee positions, including, director, NCR Corporation; director, National-American Wholesale Grocers' Association/International Foodservice Distributors Association; director, Harvard Business School Club of Houston; and trustee, Wesleyan University. He also is a member of the board of directors of Sysco.

Ceres II Discussants

Mr. David R. Beatty, President and Chief Executive Officer, Weston Foods, Ltd.

Mr. Max Downham, Vice President, Strategy and Development, The NutraSweet Company

Mr. Robert E. Flynn, Chairman of the Board and Chief Executive Officer, The NutraSweet Company

Mr. Blaine R. Hess, President and Chief Executive Officer, Thomas J. Lipton Co.

Mr. David H. Hettinga, Vice President, Research Technology and Engineering, Land O' Lakes, Inc.

Mr. Jules Hirsch, Sherman Fairchild Professor, The Rockefeller University, and Senior Physician, Rockefeller Hospital

Mr. Ralph Hofstad, former Chief Executive Officer, Land O' Lakes, Inc.

Mr. Thomas A. MacMurray, Vice President, Corporate Technical Department, H. J. Heinz Co.

Mr. Jean-Pierre Mareschi, Director, International Affairs, B.S.N.

Mr. Gerald D. Murphy, Chief Executive Officer, ERLY Industries

Mr. Gary Schwammlein, Vice President, Food Ingredients Sales, The NutraSweet Company

Mr. Robert Shapiro, Chief Executive Office, Monsanto Agriculture Co.

Dr. Robert E. Smith, Senior Vice President, Research and Development, Nabisco Bisquit Co.

Mr. Katsuhiro Utada, Honorary Chairman, Ajinomoto Company, Inc.

Mr. M.B. van Zwanenberg, Development Director, Marx and Spencer, P.L.C.

Mr. Gregory A. Vaut, Foundation of the Development of Polish Agriculture

Dr. Catherine E. Woteki, Director, Food and Nutrition Board, National Institute of Medicine

Index